Effects and Risks of Climate Change

Effects and Risks of Climate Change

Edited by **Daisy Mathews**

New York

Published by Callisto Reference,
106 Park Avenue, Suite 200,
New York, NY 10016, USA
www.callistoreference.com

Effects and Risks of Climate Change
Edited by Daisy Mathews

International Standard Book Number: 978-1-63239-167-4 (Hardback)

Contents

Preface

This book aims to highlight the current researches and provides a platform to further the scope of innovations in this area. This book is a product of the combined efforts of many researchers and scientists, after going through thorough studies and analysis from different parts of the world. The objective of this book is to provide the readers with the latest information of the field.

Climate change is a significant and lasting change in the statistical distribution of weather patterns over periods ranging from decades to millions of years. The earth is the only planet where life exists. The evolution of human life was only possible due to its favorable environmental conditions i.e. availability of water, oxygen and suitable temperature. Hence, the field of climate control requires special attention to control the rise in atmospheric temperature by controlling the emission of greenhouse gases. There is a need to conserve natural resources and look for some alternative to these resources. The book provides basic information and facts about climate change and hydrological effects, related to melting ice caps and glaciers, rise in sea level, rise in temperature and increasing number of disasters. It also discusses latest technologies applied and adopted to reduce the impact of these changes.

I would like to express my sincere thanks to the authors for their dedicated efforts in the completion of this book. I acknowledge the efforts of the publisher for providing constant support. Lastly, I would like to thank my family for their support in all academic endeavors.

Editor

Climate Change and Its Effects on Soil and Agriculture

Climate Change Adaptation Strategies in Sub-Saharan Africa: Foundations for the Future

P. J. M. Cooper, R. D. Stern, M. Noguer and J. M. Gathenya

Additional information is available at the end of the chapter

1. Introduction

Many institutions across sub-Saharan Africa (SSA) and many funding agencies that support them are currently engaged in initiatives that are targeted towards adapting rainfed agriculture to climate change. This does, however, present some very real and complex research and policy challenges. Given to date the generally low impact of agricultural research across SSA on improving the welfare of rainfed farmers under *current* climatic conditions, a comprehensive strategy is required if the considerably more complex challenge of adapting agriculture to *future* climate change is to bear fruit. In articulating such a strategy, it is useful to consider the criteria by which current successful initiatives should be judged.

Ultimately, but possibly beyond the time scale within which funding agencies will specifically support climate change research in SSA, success will be measured by clear evidence that farmers are better able to cope with current climate-induced risk and adapt to future climate change as the need for the latter becomes imperative. However, for that to happen and for agricultural research to have made a significant contribution, in the shorter term there are key 'foundation stones' that must be in place upon which such research must be built. *It is the degree to which the support provided under current initiatives is able to help lay those foundation stones that success should be judged.*

In Section 2, a brief assessment of the complexities and challenges that face agricultural research is provided and then Section 3 describes the research approach for climate change adaptation for those challenges to be successfully addressed. Section 4 illustrates the key aspects of the foundation stones that need to be in place and by which successful support could be judged. The foundation stones have been grouped under three general headings: (i) Improved access to information (ii) Enhanced research capacity and (iii) Enhancing the impact of research. In the final section (Section 5), several recommendations have been suggested for actions that funding agencies and research institutions alike might wish to consider.

2. Rainfed agricultural development in Africa: A complex challenge

Rainfed agriculture in SSA has evolved gradually over the years in response to spatially very variable environmental conditions (principally rainfall, temperature and soil types) as well as diverse cultural, social and economic drivers. As a result, the types of farming systems that we see today are also very diverse, as are the development problems that they pose. The outcome has been that agricultural development, even in the absence of projected climate change, has already proved to be a challenging undertaking and after many decades of endeavour only moderate success has been observed.

In spite of the economic importance of rainfed agriculture in the region, both in terms of its contribution to National GDP and its role in providing a livelihood to a very high percentage of the human population, investment in this vital production system, and hence its productivity, has stagnated compared with other regions of the world where small-scale rainfed agriculture is important. As population continues to grow worldwide, increased food demand, hence a need for higher crop yields, has led to more intensive land use resulting in nutrient mining and degraded soils. Such nutrient depletion on a continent wide scale can only be reversed with the help of chemical fertilizers and hence fertilizer use trends can act as a useful proxy indicator of investment in agriculture. Average rates of fertilizer use have risen ten-fold, from 5 to 50 kg ha[-1] in many parts of Asia and Latin America during the last 50 years whilst in SSA they have stagnated at a very low level of about 5 kg ha[-1] from about 1980 onwards (Figure 1).

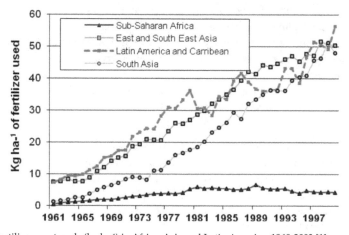

Figure 1. Fertilizer use trends (kg ha[-1]) in Africa, Asia and Latin America: 1960-2003 [1]

There are many complex and interrelated issues that contribute to this state of affairs. The outcomes of lack of investment and low production of rainfed agriculture reinforce each other, leading to poverty traps and increased vulnerability of livelihoods to climatic and other shocks [2]. This has become well recognized and an emerging political will, both within and outside SSA, to support increased investment in rainfed agriculture is gaining momentum [3]. Nevertheless, for such investment strategies to produce the needed impact on a wide scale, favourable policies, institutional arrangements and basic development infrastructure are required for proper functioning of markets. An enabling investment policy environment would therefore include the existence of proper incentives, market access, information, input supply systems and the institutions required to reinforce their use [4, 5]. However, in many countries in SSA, low per capita incomes, debt servicing and negative balance of payments at the national level still undermine both the ability of governments to invest in basic infrastructure needed for markets and the private sector to operate effectively.

However, in spite of the emerging signs of greater commitment to rainfed agriculture in SSA, there is one fundamental characteristic of this sector that will continue to pose challenges. Rainfall variability, both within and between seasons, creates an underlying climate-induced risk and uncertainty for current farm-level production as well as for the potential impact of innovations designed to improve crop, soil and livestock management practices. This uncertainty discourages the adoption of improved farming practices and the beneficial 'investment' decisions required, not only from farming communities, but also from a wide range of additional agricultural stakeholders. They show an understandable reluctance to invest in potentially more sustainable, productive and economically rewarding practices when the returns to investment appear so unpredictable from season to season [6].

Overlaid on this challenging state of affairs is the accepted prediction that, whatever happens to future greenhouse gas emissions, we are now locked into global warming and inevitable changes to climatic patterns which are likely to exacerbate existing rainfall variability in SSA and further increase the frequency of climatic extremes [7]. *'Adaptation to climate change is therefore no longer a secondary and long-term response option only to be considered as a last resort. It is now prevalent and imperative, and for those communities already vulnerable to the impacts of present day climatic hazards, an urgent imperative'* [8]. This is the challenge that many institutions, and the funding agencies who support them, are attempting to address, namely to identify, investigate and promote climate change adaptation strategies for the diverse farming systems of SSA.

In undertaking research on strategies for adapting agriculture to progressive climate change, there is the need to understand the possible nature and timescale of those evolving climatic conditions to which farmers will need to adapt to. It is also important that the uncertainties associated with these projections are understood. Uncertainty in climate projections occurs from three principal sources:

i.　Natural internal variability of the climate system – arising from factors such as variations in the therrmohaline circulation, El Niño-Southern Oscillation, and changes in the ocean heat content. These are the natural internal processes within the climate system.

ii.   Model uncertainty – climate scientists use climate models to project plausible future climate scenarios. These models are a physical representation of how the climate system works but there are still limitations in the knowledge of the process that govern the climate system coupled with limited computing resources.

iii.   Scenario uncertainty – climate projections are based on emission scenarios, such as emissions of CO_2 and other 'greenhouse' gases. These scenarios are created based on assumptions on how the future will evolve, and of course there is considerable uncertainty associated with those assumptions.

The climate community is working to reduce these uncertainties [9, 10] through, for example, using multi-model ensembles. From these uncertainties, for seasonal precipitation, internal variability is the dominant source of uncertainty for the first few decades, while model uncertainty is the dominant source of uncertainty for longer lead times. *At the moment, adaptation decisions will need to be made in the context of high uncertainty concerning regional changes in precipitation.*

To try to account for these uncertainties the IPCC, in its latest report [7], assessed results from a range of Atmosphere-Ocean General Circulation Models (AOGCM) and provided climate projections for the end of the 21st Century. Projections from all these models show substantial agreement, but as might be expected, there are still considerable differences between the various models. For example, Table 1 provides information for East Africa generated from a set of 21 AOGCMs for one of the SRES emission scenarios group (the A1B scenario) focusing on the change in climate between the 1980 to 1999 period in the 20th century integrations and the 2080 to 2090 period of A1B. Table 1 shows the minimum, maximum, median (50%), and 25 and 75% quartile values among the 21 models, for temperature (°C) and precipitation (%) change.

Month	Temperature Response (°C)					Precipitation Response (%)				
	Min	25	50	75	Max	Min	25	50	75	Max
DJF	2.0	2.6	3.1	3.4	4.2	-3	6	13	16	33
MAM	1.7	2.7	3.2	3.5	4.5	-9	2	6	9	20
JJA	1.6	2.7	3.4	3.6	4.7	-18	-2	4	7	16
SON	1.9	2.6	3.1	3.6	4.3	-10	3	7	13	38

Table 1. Projected climate change in East Africa by the end of the 21st century [7]

With regard to temperature there is a clear consensus across all models that temperatures *will* increase, although the predicted range is quite large and, agriculturally speaking, very important. A rise in mean temperatures of over 4°C by 2100 would have very different and much more dramatic impacts than an increase of less than 2°C. The clear consensus that we *are* living in a warming world has been widely confirmed through trend analyses of long-term historical daily temperature records both worldwide and in SSA. This is useful in that it provides a clear framework within which research into adaptation strategies to deal with increasing temperatures can be framed with some degree of confidence.

For rainfall, the picture is less clear with some models, for example, projecting quite severe drying in Eastern Africa in different three monthly periods and others projecting substantial wetting. However, monthly periods in which the middle half (25–75% quartiles) of the model projection distribution is of the same sign are coloured red in Table 1 and point to an emerging consensus that for both the short rains (OND) and the long rains (MAM) in East Africa, the probability is that it will become wetter over time due to an increased frequency of El Niño type events.

However, unlike the confirmation provided by trend analyses of historical weather data for temperature increases, corresponding analyses for changes in rainfall totals and distribution patterns do not currently confirm such a wetting trend in East Africa and complimentary evidence from recent publications on the impact of climate change on food security is often conflicting, for example:

- In [11] it is concluded that Eastern Africa is 'largely insulated' from the impacts of climate change with temperatures only rising by 1°C by 2030 and in general, rainfall increasing by 7 to 9% with a corresponding increase in the length of the growing season in many parts [12].
- In contrast, in an analysis of East African rainfall data, [13] suggest a decline in the long rains (MAM) in Eastern Africa which they attribute to increases in the Indian Ocean sea surface temperature, thus threatening future food security, but …
- A trend analyses of the long rains precipitation covering the same time period undertaken for five locations in Kenya [14] and one location in Uganda [15] failed to identify any such drying or indeed wetting trends.

A further additional source of uncertainty is introduced when downscaling techniques are used to try to give "better" regional projection of climate, in other words information below the grid scale of the AOGCM. These techniques include the use of sophisticated statistical methods, such as regression type models and weather generators, and also dynamical based methods such as high resolution climate models to represent global or regional sub-domains. These techniques have been used with varied degree of success. For example, [7], Chapter 11 assesses these approaches and conclude that "*downscaling methods have matured since the Third Assessment Report* [16] *and have been more widely applied, although only in some regions has large-scale coordination of multi-model downscaling of climate change simulations been achieved.*" In [17], a review of downscaling methodologies for Africa Climate applications was presented. They concluded that "*downscaling is best understood as an attempt to increase the understanding of climate change influences at the regional scale. In that context, a variety of methodologies should be explored, using all tools possible to increase that understanding*"

In summary therefore, from the outset, researchers are faced with the following challenges:

- To date, even in the absence of climate change, the development of the diverse rainfed agricultural systems in Africa has been disappointing. The low adoption of innovative farming practice has been constrained, not only by the lack of enabling policies and infrastructure, but also by the current large season to season and within season variability of rainfall distribution and the resultant risk that it poses for farm level performance and hence for returns to investment in this sector.

- Regional level projections from AOGCMs have provided a strong consensus that temperatures will increase considerably by the end of this century and this is reflected in the analyses of long-term historical temperature records both globally and within SSA.
- Similar AOGCM projections for rainfall changes are not so clear but consensus pictures are emerging. For example, in Eastern Africa, a consensus has emerged that it will become wetter in both the short and long rainy seasons. However, unlike with temperature, these wetting trends are not yet evident in the analyses of long-term historical rainfall data and published evidence of such trends is both scarce and can be contradictory.
- The use of downscaling approaches adds another layer of uncertainty to future climate projections.

3. Meeting the research challenge: a two-pronged approach

Given the constraints noted above of (i) current climate-induced risk and (ii) the predicted (although uncertain) future change in the nature of that risk due to climate change, it is now widely accepted that a two-pronged approach, sometimes referred to as the 'twin pillars' of adaptation to climate change, is needed [18, 19, 20, 6]. Such an approach recognizes that both short and medium to long-term strategies are required:

- **Coping Strategies** are those that have evolved over time through farmers' long experience in dealing with the current known and understood natural variation in weather that they expect both within and between seasons, whereas:
- **Adaptation Strategies** are longer-term (beyond a single rainfall season) strategies that will be needed for farmers to respond to a new set of evolving climatic conditions that they have not previously experienced.

In undertaking research to elucidate farmers' possible adaptation strategies, such strategies are often confused with farmers' traditional coping strategies. In the context of addressing climate-induced risk more generally, research on both is useful, but the confusion between coping and adaptation inevitably devalues the research and could well lead to erroneous recommendations.

3.1. Short-term: Coping better with current weather-induced risk to farm production

Firstly in the shorter term, since rainfed farmers are already vulnerable to current weather variability and associated shocks, it is essential to help them to build their livelihood resilience through coping better with current climate-induced risk as a pre-requisite to adapting to future climate change. Not only will greater resilience, and hence a greater adaptive capacity, allow farmers a wider range of adaptation options in the future, but perhaps more important is the consideration of the already substantial current season-to-season weather ranges and the extent to which these ranges will, or will not change in the future. Whilst temperatures are already increasing and changes in rainfall amounts and patterns may begin to become clearer in the future, the question remains to what extent will farmers experience conditions under progressive climate change that they are not already experiencing today?

In [21], an example based on an analysis of the impact of an assumed 3°C rise in temperature on the length of growing period (LGP) at a semi-arid location in Kenya (Figure 2) is provided. Whilst a possible 3°C increase in temperature in the future reduced the mean LGP by about 8% across the 45 years analyzed, the projected LGPs ranged from 63 to 152 days compared with the 76 to 175 experienced today. In other words, in about 80% of the seasons that would occur in the future under the assumed 3°C temperature increase, farmers would *still* be experiencing similar LGPs to those that they are already experiencing today.

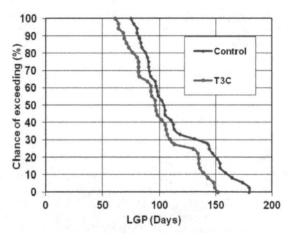

Figure 2. The effect of a 3°C rise in temperature on the Length of Growing Period (LGP) at Makindu, Kenya. 1959-2004 [21]

In conclusion, helping farmers cope better with current climate variability is a win-win situation. It will improve their lives now and help in the future.

3.2. Medium to longer-term: Adapting agriculture to future climate change

In the medium to longer term and as climate change becomes more obvious, both in its identification and impact, farmers will have to adapt their farming practices to a new set of weather-induced risks and opportunities. However, there are three major complicating factors in this second aspect of the strategy:

i.   We have already referred to the uncertainty associated with future climate projections and the new nature of the climate that farmers will need to be adapting to. In other words, the 'climate change goal posts' for which research should be aiming are uncertain. That in itself poses considerable research questions.

ii. This is further complicated by the fact that climate change will be progressive over time and the length of time it will take before a *'final climate state'* is achieved (if indeed it is achieved) is unknown. In other words – the goal posts will be continually moving with time! This infers that climate change adaptation research itself has to be ongoing and continually self-assessing.

iii. We have already mentioned the natural and characteristic variability of rainfall in the region, both within and between seasons. Drying and wetting cycles are a natural characteristic of this variability and can be quite lengthy as at Bulawayo in Zimbabwe (Figure 3), but usually are somewhat shorter, 3 to 4 years in duration (see Figure 6). This characteristic of the climate itself imposes the caveat that one should be careful not to mistake such relatively short term trends for long-term climate change. But that in itself begs the question *"how does one know whether a currently observed trend is short term or the start of longer term climate change?"* In Kenya for example, there have recently been a series of drier than normal seasons. Is this just another short term drying trend as has been observed before or the start of a longer process of climate change as many currently seem to assume?

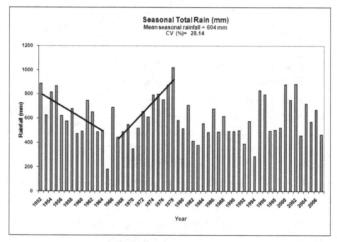

Figure 3. Relatively long term (10 year) drying and wetting cycles at Bulawayo, Zimbabwe (1952-2007).

4. Supporting evidence-based climate adaptation strategies: The foundation stones of success

In spite of the challenges illustrated in Section 2 and the complexities of short-term versus medium to long-term approach to adaptation highlighted in Section 3, the agricultural research community does urgently need to initiate research now to identify medium to longer term climate adaptation options for the future. *What are the key elements that need to be put in place for the research community in Africa to undertake research into climate adaptation and be able to provide evidence-based adaptation options to farmers?*

The observations and conclusions described in this study are not the outputs of a formalized research approach or specific experimental design. They are based on the long-term, substantive and collective experience of the authors who have interacted for many years with a wide range of projects addressing the challenges of both current climatic variability as well as future climate change impacts on rainfed agricultural production and farmer livelihood. Such interactions have arisen through a range of circumstances from (i) living and working at agricultural institutions based in SSA, (ii) working for advanced research institutions that have partnered with African institutions addressing climate variability and change impacts to (iii) providing extensive training and support for African researchers to enable them to acquire the scientific skills required to undertake such research.

Most recently, since 2010, the authors have worked together as a team within a project supported by the Rockefeller Foundation through a grant to the Walker Institute for Climate System Research at the University of Reading, UK, entitled *'Supporting the Rockefeller Foundation Climate Change Units (CCU) in East and Central Africa'*. As part of that grant, Rockefeller asked the authors to address the question of *'What would success look like?'* with regard to their support to eight CCUs in Tanzania, Kenya, Rwanda, Uganda and Ethiopia.

The authors highlight here what they believe are the most important 'foundation stones' upon which rigorous and useful climate change research should be built, and in this context, research that addresses both aspects of the 'twin pillars of adaptation' to climate change, as discussed in Section 3, is included. These foundation stones are summarised in Table 2 according to the three groupings of (i) Improved access to information, (ii) Enhanced research capacity and (iii) Enhancing the impacts of research.

Improved access to information	Enhanced research capacity	Enhancing the impacts of research
Access to literature and a database of projects	*Developing conceptual frameworks for the impact pathways of change*	*Producing quality **written** publications to influence stakeholders*
Access to historical climate data	*Risk and trend analyses of historical weather data*	*Producing **visual** presentations to influence audiences*
Access to up-to-date curriculum in Universities	*Analyses of impacts of climate variability and change on agricultural production*	*Archiving primary data in accessible formats*

Table 2. Foundation stones upon which rigorous climate change research should be built.

4.1. Improved access to information

Most of the scientists currently engaged in climate change research in the area have been trained in various disciplines of agriculture or the social sciences but do not have a fundamental knowledge of the climate sciences. Yet the onus is upon them to plan and execute relevant and innovative research targeted towards helping farmers cope better with current climate-induced risk and to adapt to climate change. For them to successfully address this challenge, it is imperative that they have easy access to:

- Published and unpublished information contained in the literature that is related to their field of enquiry.
- Information about both completed and on-going projects addressing climate risk management and adaptation to climate change in SSA, and
- Long-term daily weather datasets collected from recording stations close to the location where their studies will be based, without which 'hard' climate risk and climate change research is very difficult, or indeed, near impossible.

Without access to such information there is a real danger that the research undertaken will fail to be well prioritized, rigorous, relevant and non-repetitive of that already undertaken or currently on-going.

In addition, it is equally, if not more important that the next generation of scientists who undertake such research will have had a much more substantive exposure to climate science in their graduate and postgraduate training than is currently the case. This will require the building of specific and up-to-date teaching material into their curricula.

4.1.1. Access to literature and a database of projects

Over the last four to five decades, a great deal of research has been conducted in East Africa and across SSA. Much of this is relevant to the first of the twin pillars of adaptation, namely that of helping farmers cope better with current climate-induced risk. Indeed, research that has specifically studied farmers' weather driven coping strategies has been on-going in Africa for twenty or more years [22]. More recently, and of direct relevance to the second of the twin pillars, namely helping farmers to adapt to climate change, much has also been reported on climate change projections in SSA and their possible impacts on rainfed agriculture and food security.

In East Africa for example, the impacts of climate change projections on the length of the growing season have been examined by [12] and more recently, on food security, by [11] and have also been reviewed by [23]. In [24] a regional climate model to compare the effects of projected greenhouse gases concentration and land use land cover change on spatial variation of crop yields in East Africa has been used. In addition, the extent to which projected rainfall and temperature trends are, or are not, yet becoming evident in daily and seasonal weather observations have also been assessed [13, 15, 14].

The authors have found that when supporting scientists in developing their research proposals to address current climate risk and future climate change, almost all the them had trouble in accessing published information that is relevant to their proposed research. In particular, it is apparent that many do not have access to the literature reporting climate change projections for their region [7] or individual papers assessed by the IPCC. Hence they undertake their research in the absence of an understanding of the emerging consensus as to likely future climates in their regions to which farming families will have to adapt.

Such projections can be obtained as coarse resolution outputs of the global climate models (GCMs) from the Intergovernmental Panel on Climate Change (IPCC) data portal [25] and finer resolution downscaled data can be downloaded from the Decision and Policy Analysis Program of the International Centre for Tropical Agriculture (CIAT) website [26].

However, in spite of the relatively easy access to available literature, it remains quite common for the research problem to be defined and the objectives and activities detailed with very little, or indeed any, literature review having been undertaken. In such cases, 'undertaking a literature review' is often stated as one of the objectives or activities. Clearly, it is not ideal to first prescribe the research to be undertaken and then undertake a literature review to find out what has already been done and what is already known.

In some respects, this lack of access to published material is understandable, but in the age of the web and Google Scholar, much more literature searching should have been possible. In addition, SSA is fortunate to have many international research bodies with their offices based in African countries. For example in East Africa, the Office of the IGAD Climate Prediction and Applications Centre (ICPAC) are in Nairobi, Kenya as well as several relevant international centres (e.g. ILRI, ICRAF, CIMMYT, ICRISAT, CIAT, and CABI). All these centres are involved in climate risk management research and have much published and unpublished information available which can be easily accessed through a visit. In addition, many of the African universities and research institutions have well stocked libraries which can also be visited. However, this could be difficult for many scientists who are based some distance from a research headquarters or university or do not have easy access to the internet. *In that respect, a dedicated review of the literature and the production of an easily accessible annotated bibliography of up-to-date and key climate risk management and adaptation literature would be very helpful to many scientists.*

Finally, there are a large number of completed and ongoing climate related research initiatives, many of which are supported with external funding. Such initiatives have an enormous range of scales at which they operate. Whilst some are quite modest in both scale and funding support, others are very large, region wide in scope and long-term in duration. The Climate Change, Agriculture and Food Security (CCAFS) [27] initiative, active in both West and East Africa is probably the single biggest example. The CCAFS Research Programme is founded on the principle of sharing of methods, data and publications. What however is of great concern is the lack of information, or at least the sharing of information, concerning the existence, scope, activities and results obtained by past and on-going projects that are addressing climate risk and adaptation to climate change. This problem appears to exist across regions, countries, institutions, funding agencies and even within institutions themselves. Indeed, within the Rockefeller initiative, several grant recipients have found it necessary to produce a database of climate related projects within their own institution in order to better understand the scope of their own past and on-going climate related research.

Given the real danger of precious financial resources being used to support research that has already been undertaken, or is on-going elsewhere in SSA, it is very important that, at least on a regional basis, the information held by individual institutions and funding agencies is collated, made easily accessible and shared more widely.

4.1.2. Access to historical climate data

A great deal of daily weather data has been collected in each country of SSA over the last 60 to 70 years and indeed some records go back more than a 100 years [28]. The network of recording stations is not very dense but this information is fundamental to climate risk and climate change research. Easy access to high quality long-term daily weather records is essential to:

- Undertake detailed risk analyses of important weather events that affect crop and livestock performance and assess to what extent there are or are not significant trends in those parameters that could be ascribed to climate change [29].
- Compare widely held perceptions of climate change, often documented in farmer surveys, with the realities of long-term daily weather data analyses [14,15].
- Drive a range of hydrological models as in [30] and crop growth simulation models as in [21,31] which integrate the impacts of rainfall, temperature and solar radiation variability into expressions of agricultural and environmental risk response.

Whilst it is true that a new generation of weather generators (for example MarkSim and Weatherman) can produce long-term weather datasets that are mathematically and statistically representative of real time weather data for any given location and can be used effectively to drive agricultural models (e.g. [31]), they can only ever be a first approximation for estimating climate-induced production risk and cannot be used to assess possible trends in important climatic parameters.

However, in spite of the evident importance of national scientists being allowed easy access to such national datasets, this is not the case in most countries in SSA where the National Meteorological Services (NMS) require payment for the datasets and even when this can be afforded, the data provided is often of poor quality and limited duration.

This constraint is well known and many individuals and institutions are now engaging with the NMS. In the context of climate-induced risk in the vital sector of rainfed agriculture and the possible impacts of climate change in exacerbating those risks, such data are too important to be considered the property of a single entity and should be declared national public goods. Indeed, [32] concluded a seminal review of the potential of seasonal climate forecasting in SSA with the recommendation *"Finally, meteorological data [in sub-Saharan Africa] should be treated by national policy as a free public good and a resource for sustainable development across sectors"*.

Engagement with the NMS by funding agencies, influential scientists and global organizations such as the World Meteorological Organization (WMO) would pay dividends. While it is clear that NMSs need to cover their costs of data collection, they should also be able to position themselves to generate even greater resources from offering innovative, demand-driven, value added weather products to the agricultural community.

However, it must also be recognized that the sharing of raw data in collaborative partnerships between universities, National Agricultural Research Institutions and NMS must be a "two way street". It is important that credit be given to the NMS for the considerable efforts they make to computerise and quality-control their historical records. Were they to agree to an exchange of their records for similar long records of yield and growth data from agricultural research scientists, they would probably find many researchers equally unwilling to share their own data. Sometimes such sharing would expose the poor strategies for quality control of the agricultural datasets and it might even be that they are unavailable, since a great deal of data remain with students, or staff who have left the institute. And this is often despite the use of public funds for the research.

Since the 1980s, and led partly by the World Meteorological Organisation (WMO), the climatic community has devoted considerable resources to the management and quality control of the long-term historical records. This has yet to be matched by the agricultural research initiatives. It is important that funding agencies who support climate adaptation research help to ensure that primary datasets that are developed using their funds are well organised during the research period and made publically available once the researchers have completed their work.

4.1.3. Access to up-to-date curriculum in universities

The evidence from analyses of long-term temperature data is unequivocal in demonstrating that the world is warming at an unprecedented rate, and whilst parallel analyses of rainfall records as yet seldom show significant trends, there is little doubt that such trends will become progressively apparent in the coming years. Simply put, over time the impacts of climate change will increasingly be felt in the agricultural sector and indeed across all sectors in SSA. In response to this, it is important that the next generation of agricultural scientists, both undergraduates and postgraduates, have a far stronger grounding in the climate sciences and a far greater understanding of the interactions between climate, people and agriculture than is currently the case.

Globally, many universities are offering appropriate courses at the graduate and postgraduate level and the opportunity for African students to register for such courses will remain available providing the funding is available to support them. However, the number of students who can benefit this way will always be limited and is unlikely to create the critical mass of expertise that will be required in the future. What is needed is the universities of SSA themselves to develop well structured, targeted and up-to-date curricula. To some extent, this can be achieved by individual universities assembling existing and relevant modules from a range of different degree courses and building them into a more dedicated and structured course. However, this approach is unlikely to provide an optimal solution for two reasons. Firstly, *within* any given university, existing modules are unlikely to cover the full scope of the climate change agenda that is required and the modules that do exist will possibly be of different quality. Secondly such modules were probably designed to address the needs of specific courses rather than to be integrated into a comprehensive teaching agenda addressing agriculture, climate risk and climate change.

A better solution would be achieved through greater interaction and discussion *between* universities in SSA to (i) agree on the essential topics that would need to be included in comprehensive graduate and postgraduate climate related courses, (ii) assess what material is currently "at hand" both within and from outside SSA and (iii) build the best of that material into a comprehensive and integrated agenda of adaption to and mitigation of climate change. In addition, universities would need to asses to what extent the staff responsible for the teaching of such a course would require some specialized training in its delivery.

Training is also urgently needed for the current generation who are working in a wide range of climate related research and development projects. These include not only researchers, but also agricultural extension staff, NGOs and farmer's organizations, all of whom need to understand the 'twin pillars of adaptation' described earlier. Currently all these organizations and farmers themselves are acutely aware of the general topic of "climate-induced risk and change", but all too often they struggle to find the most useful way to respond.

Where funding agencies do support the development of new curricula together with the corresponding training material, they should also ensure that the resulting materials are "open educational resources" in order to ensure their widest possible use.

Given the great importance of establishing a critical mass of 'climate informed' research personnel for the future, it is important that initiatives to develop a comprehensive climate change curriculum are considered a priority and are provided with funding to support their initiation and implementation.

4.2. Enhanced research capacity

An increasing number of projects are being funded to undertake research that will investigate ways in which the negative impacts of climate change on rainfed agriculture in SSA can be mitigated. As mentioned in the previous section, the majority of agricultural scientists engaged in this research have little, if any, background training in climate science. Given this, it is not surprising that many are not aware of the complexities of the climate science that underpin climate change projections or the associated uncertainties and the challenges inherent in climate change adaptation.

None of this is surprising, nor is it unique to scientists in SSA, but it does reinforce the idea that current initiatives aimed at improving the climate change curricula has a very high priority as has the continued capacity enhancement of those who are already involved in such research. Through visiting and working together with the researchers in East Africa over the last 18 months, the authors have identified three areas where immediate priority should be given in capacity development, namely:

- Developing conceptual frameworks for the impact pathways of change
- Risk and trend analyses of historical weather data
- Analyses of impacts of climate variability and projected climate changes on agricultural production.

4.2.1. Developing conceptual frameworks for the impact pathways of change

Over the last decade, climate change has become one of the hottest topics across SSA. It is being discussed across all levels of society from high-level government representatives, representatives of national and regional bodies, to research and development agencies and through to private citizens and to small-scale farmers.

However, investigations into the possible current and future impacts of climate change need to be put into the context of the well documented and on-going impacts of other drivers of change, some of which are currently likely to be of more immediate importance than climate change itself.

Without doubt, population growth is the single greatest primary driver of change in many parts of the world and particularly in SSA. Indeed, globally, it is the primary driver behind increased greenhouse gas emissions and hence climate change.

In East Africa for example (Table 3), between 1961 and 2011, the population (Ethiopia and Somalia excluded[1]) has risen from 49 to 208 million people, and is projected to rise to over 500 million by 2050. The direct impact of this is most clearly illustrated by considering the number of additional people that have to be fed each year without there being the opportunity to greatly expand the area of high potential land available for agriculture. In the ten years leading up to 2011, there was a mean of (excluding Ethiopia and Somalia) 4.4 million additional mouths to feed *each year*, and this is projected to rise to a mean 8.8 million additional mouths to feed *each year* in the 20 years leading up to 2050. That in itself presents, and will continue to present, enormous challenges for policy formulation and agricultural development.

Over and above that direct consequence, there are a range of 'secondary' drivers of change that have stemmed from population growth. Whilst the impact of some of these 'secondary' drivers of change on agriculture have been positive, such as the creation of new markets, increased demand for crop and livestock products and greater opportunities for off-farm employment, almost all the impacts of population pressure on rural communities, agriculture and the environment have been negative and are likely to become worse in the future.

Therefore, it is essential to formulate a clear 'climate change' hypothesis in any research project and then situate that hypothesis in a conceptual framework that considered the

---

[1] NOTE: statistics for Ethiopia are not available for the full period, but in 2011 the population stood at 85 million and is projected to rise to 145 million by 2050. No statistics are available for Somalia.

	1961	1971	1991	2011	2031	2050
Burundi	3.0	3.6	5.7	8.6	11.6	13.7
Kenya	8.4	11.7	24.2	41.6	67.4	96.9
Madagascar	5.3	6.7	11.6	21.3	36.2	53.6
Rwanda	2.9	3.9	6.9	10.9	18.0	26.0
Sudan	11.8	15.2	27.2	44.6	68.1	91.0
Uganda	7.0	9.7	18.3	34.5	61.1	94.3
Tanzania	10.4	14.0	26.3	46.2	84.2	138.3
TOTAL	48.8	64.8	120.2	207.7	346.6	513.8
Additional people to feed *each year (millions)* (Mean for time period)	-	1.6	2.8	4.4	6.9	8.8

Table 3. Actual and *projected* human population trends in selected East African countries (millions) Source FAOSTAT

possible impact pathways of other drivers of change. For example, one could base their study on the hypothesis that 'climate change was leading to a decline in rainfall and was responsible for an increase over time of the number of people and livestock suffering food and feed shortages'. A conceptual framework that situates this hypothesis in the context of related impact pathways of human and animal population growth can be developed (Figure 4).

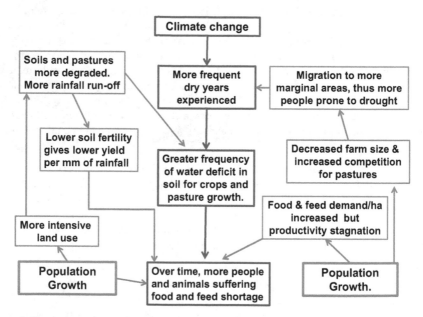

Figure 4. Illustrative impact pathways of climate change (red) and population growth (blue).

Unless researchers are trained to undertake this type of thinking process, misinterpretation of results could lead to quite serious consequences. For example, in the studies by [14, 15], survey work indicated that farmers believed that declining rainfall was responsible for lower crop yields and farm productivity. Taken at face value, such a conclusion could well have led to initiatives to identify appropriate adaptation strategies for drier conditions such as moisture conservation practices or the use of shorter duration crop varieties. This would have been a serious mistake since in each case further investigation showed that other drivers of change were responsible for declining productivity, and not declining rainfall resulting from climate change.

Putting climate change in the context of other drivers of change is essential for researchers to have confidence that the processes, impacts and research innovations under investigation are properly focussed on climate-induced risk and change and are not in response to other drivers of change.

4.2.2. Risk and trend analyses of historical weather data

Section 4.1.2 discussed the importance of access to high quality long-term daily weather data and indicated the value of such data in enabling climate risk analyses and trend analyses. Both types of analyses are important.

Risk analysis: Rainfed crop growth, development and yield formation can be greatly influenced by individual weather events or by a combination of events. Such events could include for example (a) a series of rainy days at the start of the season that enables planting, (b) a period of drought following planting which leads to seedling death, (c) very heavy rain that causes water logging and impairs root growth or signals the possible onset of root diseases such as root rot in beans, (d) below optimal rainfall or super-optimal temperatures at flowering resulting in poor yield formation or (e) terminal drought resulting in poor grain filling or the possible onset of disease outbreaks such as aflotoxin in groundnuts.

Risk analyses of long-term daily weather data can assess the probability (or risk) of such events occurring and can assess how that risk can be 'beneficially managed' through, for example, the choice of planting date or the choice of crop varieties of different maturity lengths, or even the choice of different crops. Examples of such analyses has been undertaken and published by several authors from the 1980's onwards and most recently by [29] who used eighty-nine years of daily rainfall data from Moorings recording station in Southern Zambia as a case study in which a range of agriculturally important weather events such as those above were investigated using the statistical software Instat [33] and Genstat [34] as well as a simple water balance model [35].

One example they gave addressed an issue that is always a widespread priority concern of rain-fed farmers, namely, *'when has the rainy season started and when is it 'safe' to sow their crops'?* The example given stemmed from discussions with farmers near to the recording station which had indicated that in general, they tended to plant their maize as soon after mid-November as possible, but required that at least 20mm of rain had fallen within a 3-day period before they would do so. They also indicated that if there was a 10-day dry period in the month following planting, then maize seedling death occurred, necessitating re-planting. The results of the analyses are given in Figure 5 which shows the date, according to the farmers' criteria, when the first planting took place each year and in which of those years a dry spell occurred that necessitated re-planting. There are 12 such occasions in the 88 years, indicating that the risk of not succeeding with early planting was about 14%. If the dry-spell following planting condition is changed to 12 days length, (not shown) then the risk dropped to 8 years in the 88, or about 9%. This indicates the improvement that might be achieved with a more drought tolerant crop, or with simple moisture conservation measures to reduce soil surface evaporation losses such as soil surface mulching.

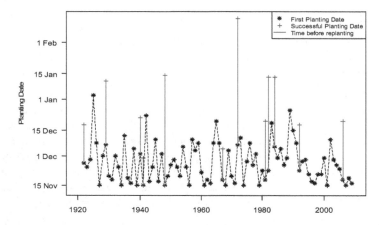

Figure 5. Date of the start of the rain, at Moorings, Zambia from 1922 to 2009 [29]

Such types of analyses provide a valuable additional source of *ex ante* risk information that will help prioritize what field-based investigations need to be undertaken.

Trend analyses: Given the uncertainties of climate change projections (see Section 2) and the fact that such uncertainties become greater as smaller time and spatial scales are desired, trend analyses of existing long-term historical climate data can be considered as '*the gold standard*' of assessing the extent of current climate change at locations where adaptation research is being undertaken [36].

It is important for two reasons:

- Firstly, using appropriate statistical curve fitting approaches to long-term data sets helps avoid the danger of mistaking short term trends of a few seasons with long-term climate change. Such cycles can be relatively long-term (see Figure 3 for Bulawayo, Zimbabwe) or shorter term as illustrated for total seasonal rainfall at Makindu in Kenya (Figure 6) where the short term wetting and drying cycles are apparent (e.g. 1963-1966, 1974-1978, 2000-2004), but fitting a line to the complete dataset showed no significant trend in either direction. This is in contrast to fitting curves to the maximum and minimum temperature data from the same location (Figure 7). Whilst the same sort of season-to-season variability in temperature is noted, fitting a curve to the complete dataset *did* show a significant increase in both maximum and minimum temperature.

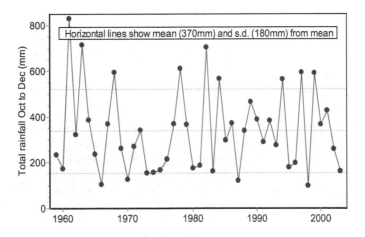

Figure 6. Seasonal (OND) total rainfall at Makindu, Kenya (1954-2004) [23]

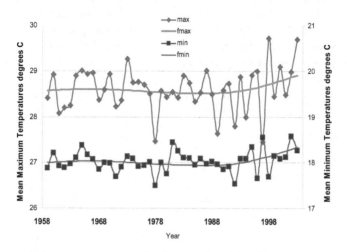

Figure 7. Seasonal (OND) mean maximum and minimum temperatures at Makindu, Kenya (1959-2004) [23]

- Secondly, a great deal of research currently underway within SSA is centred on survey work that investigates farmers' perceptions of climate risk and possible climate change and their associated coping and adaptation strategies. Having long-term weather data at hand to compare farmers' perceptions with the 'hard' risk and trend analyses of recorded weather data can be invaluable in identifying to what extent they are correct or indeed whether perhaps they are responding to other drivers of change. This is exactly what happened in the studies in semi-arid Kenya by [14]. Farmers perceived that climate change had caused declining rainfall amounts since the early 1990's and which they felt had resulted in declining maize yields. However, trend analyses of the long-term historical rainfall data from 5 locations in the study area showed no decline in rainfall amounts or changes in their distribution patterns. Further studies, whilst confirming that district level yields had indeed declined as perceived by farmers, showed that this was due to (i) a reduction in fertilizer use as a result of an increase in its price following structural readjustment during the 1990's, and (ii) migration of farmers to land with a lower yield potential due to population pressure. It was not due to climate change.

To be able to perform climate risks analysis and trend analysis of any long-term data record, the use of statistical packages is imperative for researches to be able to produce real evidence of change and hence be able to give evidence based advice.

Given the importance of this type of analysis for climate change adaptation research, the need for training on statistics in applied climatology is essential.

4.2.3. Analyses of impacts of climate variability and change on agricultural production

In addition to the type of climatic analyses described above, a further step can be taken by using a range of models to analyze the impacts of variable weather on many aspects of agricultural production, including crop, livestock, pastures and trees and shrubs using simulation models. Clearly, such models are an important tool for researchers to use if they are interested in assessing the integrated impacts of different components of climate variability (principally rainfall, temperature and solar radiation) and climate change on rainfed agricultural production.

Globally, a large number of such models have been developed. They demonstrate an equally wide range of characteristics that embrace different aspects of agriculture (i.e. crops, livestock, pastures, trees), the spatial scale at which the model operates (i.e. from plot level → farm → water catchment) and the complexity of inputs that are required to calibrate the model for the purpose the researcher has in mind. For example some models require daily weather values for rainfall, temperature and solar radiation whilst others can run with monthly means and totals. In [23] an illustrative and descriptive list of such models (and the web address where more information can be found) is provided. Because the output of such models is dependent on the climate, soil, crop, pasture and livestock management input information that the researcher provides, they can be used to complement field-based research by providing an *ex ante* evaluation of the effectiveness of any given intervention across a wider range of conditions and over a longer period of time than is usually possible through field-based research alone. This provides an additional and important source of information to that obtained from field trials when researchers are required to formulate recommendations for policy makers' consideration.

Two of the most widely used crop models are the Decision Support System for Agricultural Technology (DSSAT) and the Agricultural Production Systems Simulator (APSIM). These models *integrate* the impact of variable daily weather (principally rainfall, temperature and solar radiation) with a range of soil, water and crop management choices. Since they are 'driven' by daily weather data, they can be used to assess the impact of season-to-season climate variability on the risk associated with a range of agronomic strategies for a wide range of crops, trees and pasture that are important to African farmers. When properly calibrated they can provide an impressively accurate simulation of what occurs in 'real life'. An example of their power is illustrated in the study by [31] who used APSIM and 50 years of daily weather data from Kitale, Kenya to investigated the effect of a factorial combination of weed control (2 levels), seeding rate (6 levels) and N-fertilizer application (8 levels) on the growth and yield of maize – *equivalent to running a trial of 96 treatments for 50 years!* When compared with the data produced from an intensive agronomic investigation undertaken at Kitale nearly 40 years ago [37], the simulations produced by APSIM closely mirrored the agronomic responses observed by [31] (Figure 8). Perhaps more important is the fact that such simulations using long-term weather data can provide a more comprehensive assessment of climate-induced risks than is usually possible to achieve through field-based research which is inevitably constrained both by expense and by the length of time such studies can be continued.

Calibrating these types of models for different soil types and different crops and crop varieties is not a trivial exercise, and they do need some careful training and some follow-up support until the user becomes familiar with their use. However, once they are properly calibrated and the skills to use them are mastered, they are very powerful tools indeed for investigating climate-induced production risk associated with a broad range of possible interventions.

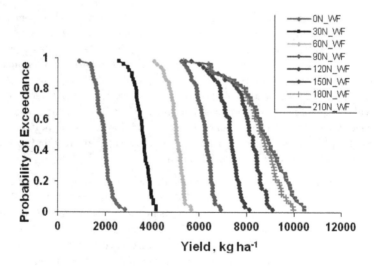

Figure 8. Probability distribution of simulated (APSIM) maize yields at different fertilizer levels (kg N ha⁻¹) in weed free (WF) maize at Kitale, Kenya [31]

Moreover and importantly, they are equally powerful in simulating the impacts of possible changes in CO_2 levels and temperature and rainfall regimes that could result from a range of climate change scenarios. Figure 9 provides an example of such an analysis where the disaggregated and aggregated impact of a climate changes [7] for Southern Africa were examined, namely:

i. CO_2 increased from 350 to 700 ppm,
ii. Temperature increased by 3°C and
iii. Rainfall decreased by 10%
iv. Combined effect of (i), (ii) & (iii)

These scenarios were compared with the baseline simulation of 'today's climate' for a well managed crop of groundnuts at Bulawayo in Zimbabwe using 56 years of daily weather data [21].

Earlier, cost was one limitation in the use of these software packages. Recently, a welcome change is that the developers of both APSIM and DSSAT have decided that researchers may have access to their software at no financial cost.

Given the value of simulation models funding agencies might consider providing dedicated funding to support learning and capacity enhancement in their use.

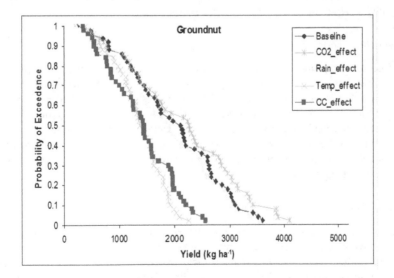

Figure 9. The probability distribution of the disaggregated impact of simulated (APSIM) climate change on the yield of groundnut compared with today's baseline climate. Bulawayo, Zimbabwe. [21]

However, organisations sometimes use training opportunities more as a way of rewarding staff, rather than to change the working practices of individuals and organizations as part of a planned capacity enhancement strategy. *It is suggested that proposals that include capacity-building should recognise that evaluation of the training components will be based on changed working practices of the trainees and of the organisation itself* [38]. The satisfaction and learning of the participants in the training courses is necessary, but is not a sufficient reason for including a training component.

4.3. Enhancing the impacts of research

If the six foundation stones discussed above under *'Improved access to information'* and *'Enhanced research capacity'* are in place within the region, then there is a strong likelihood that the quality, relevance, analyses and outputs of the work will be greatly improved. But the job does not end there. If those research results are to have the desired impacts on the welfare of rain-fed farming families, it then becomes even more important that the outputs of the research are made as visible and persuasive as possible through good reporting. Three further 'foundation stones' are needed, namely (i) Producing written publications and reports to influence stakeholders, (ii) Producing visual presentations to influence audiences and (iii) Archiving the primary data in accessible formats for further analyses.

These three particular foundation stones are general in nature and are important to all fields of research and not just to research addressing agricultural adaptation to climate change. However, funding agencies may well feel that there are actions that could be taken specifically to help those scientist that they are supporting.

4.3.1. Written reports and publications to influence stakeholders

As with all research, funding agencies would wish to see the results and conclusions of the research that they have supported properly written up and reported in an analytical, persuasive and easily understood format. These reports need to be targeted towards several audiences, ranging from published articles for the wider academic community, advisory manuals for extension agents to policy briefs for decision makers. Each of these requires different types of 'content', format and writing style. Producing these different types of reports is a skill that some people have to a greater extent than others, but fortunately it is a skill that can be taught.

4.3.2. Visual presentations to influence audiences

Increasingly, both within the region and internationally, there are opportunities for scientists involved in climate risk and adaptation research to attend meetings and conferences to give visual presentations of their research. Written work is almost always targeted towards a specific audience who will usually have time available to spend reading the report in detail. Conferences, on the other hand (i) tend to attract audiences with a wider range of specific interests, (ii) allow little time to get the message across, and (iii) usually try to schedule as many presentations as possible.

Under these circumstances, a poorly constructed and poorly delivered presentation will almost certainly fail to have any impact and will become quickly forgotten. The skills required to construct and deliver a compelling presentation are quite different from those required to write good reports; but again, they are skills that can be taught.

If individual institutes do not have access to communication experts who have the skills to help in written and visual presentation, then the training of scientist themselves in these skills might well be a useful activity to be considered for targeted funding support.

4.3.3. Archiving primary data in accessible formats

Funding agencies could play a key role in promoting the archiving of the raw research data that their funds have been used to produce. This applies to all fields of research, but it is particularly appropriate to consider this role in research linked to climate variability and change. This is because these projects often include a demand for the archived primary climatic data from the National Met Services (NMSs) who have themselves devoted considerable resources to the archiving of their historical data. There may sometimes be problems in gaining access to these data, but at least they exist. This is not the case with most raw research data collected by research programmes or by universities. It is time that this changed.

There are both practical and moral issues to be resolved before research scientists will agree to archive their raw data. The key moral issue is that of ownership. Do the data belong to the funding agency, to the individual scientist who was responsible for the data collection, or to the organization that employs the scientist? Failure to resolve these issues limits much current research. Indeed, on occasions scientists even fail to share their data between members of their research team.

The practical issue is how the raw research data should be archived, if there is agreement on the moral issues, in particular on the data ownership.

The reason why it is highly appropriate for funding agencies to be involved in this topic now is twofold:

- Solutions to the practical problem exist. For example, the Dataverse Network at Harvard University [39] is an online digital repository for research data. Although this is initially promoted as being for the social sciences, it also provides a potential solution for all research data collected by different projects. Potential users may either archive their data and supporting documents on the Harvard site, at no cost, or they may choose to download the archiving software (also at no cost) and create their own local "Dataverse".
- Funding agencies are well placed to help recipient organisations to address the moral issues because:
  - They have provided the funds; however
  - They do not want any of these data for themselves, i.e. they keep the moral "high-ground". They merely want to ensure that the data remain available for the benefits of all partners and of future research activities.

In addressing these issues it is vital that the topic of ownership, and hence of archiving, is addressed at the start of any project. Thus, when researchers agree to undertake a project, they also agree to the archiving of their data and of supporting documents.

5. Conclusions

To ensure that African researches are equipped with the best tools and develop strong research capabilities, the authors believe that there is an urgent need to invest in: (i) ways to improve access to information, (ii) ways to enhance the research capacity, and (iii) ways to enhance the impact of the research undertaken.

The following recommendations are the result of extensive work that the authors have undertaken in SSA to support researches to develop adaptation strategies that build resilience of agriculture to climate change. These recommendations are summarised here:

Improved access to information

- Support for a dedicated review of the literature and the production of an easily accessible annotated bibliography of up-to-date climate risk management and adaptation literature.
- Development of 'databases' of previous or on-going projects held by individual institutions and funding agencies.
- Access to high quality and long-term daily weather data is crucial for a critical analysis of climate-induced risk and climate change research. Such data is too important to be considered the property of a single institution and should be viewed as a public good.
- Given the importance of establishing a critical mass of research personnel for the future, support for the development of a comprehensive and integrated curriculum on climate change is essential.

Enhanced research capacity

- Promote rigorous scientific research design and approaches to make sure that research on climate change adaptation is conceived and undertaken in the context of the impacts of other important drivers of change.
- Support for training, both in climate risk analysis and in the use of weather-driven crop growth simulation models, to assess the impact of climate variability and change on agricultural production.
- Individual training should not be funded in isolation, but rather as an agreed capacity development program for institutes to develop the improved working practices of their staff and of their institute in this complex area of climate variability and change.

Enhancing the impacts of research

- Support for good reporting, both written and visual, is essential as it ensures a higher and wider impact of the research outputs and consequently better outcomes and decisions.

- Ensure that primary datasets that, developed using public funds, are properly archived and are made publically available once the researchers have completed their analyses and reporting.

Author details

P. J. M. Cooper*
School of Agriculture, Policy and Development, University of Reading, UK

R. D. Stern
Statistical Services Centre, University of Reading, UK

M. Noguer and J. M. Gathenya
Walker Institute for Climate System Research, University of Reading, UK

Acknowledgement

Authors gratefully would like to acknowledge all funding agencies that have supported the work. Thanks are due to the African Development Bank who supported a four year project (2007 – 2010) entitled *'Managing Uncertainty: Innovation Systems for Coping with Climate Variability and Change'* through a grant to the International Crops Research Institute for the Semi-arid Tropics (ICRISAT), and to the Rockefeller Foundation who are currently funding the authors' project entitled *'Supporting the Rockefeller Foundation Climate Change Units (CCU) in East and Central Africa'* (2010-2014).

6. References

[1] FAO. 2005. Increasing Fertilizer Use and Farmer Access in Sub-Saharan Africa: A Literature Review. Agricultural Management, Marketing and Finance Service (AGSF), Agricultural Support Systems Division. Rome: FAO

[2] World Bank (2000). *Can African Claim the 21st century? The World Bank*, Washington, D.C. World Bank

[3] Sanchez, P.A., Denning, G.L. and Nziguheba, G. (2009). The African Green Revolution moves forward. *Food Security*1:37–44.

[4] Barrett, C., Lynam, J. and Place, F. (2002). Towards improved natural resource management in African agriculture.In *Natural Resources Management in African Agriculture: Understanding and Improving Current Practices.* (EdsC.B. Barrett and F. Place). Wallingford, UK: CABI Publishing, 287–296.

[5] Collier, P. and Gunning, J. (1999). Why has Africa grown slowly? *Journal of Economic Perspectives* 13: 3–22.

[6] Cooper, P. J. M., Dimes, J., Rao, K. P. C., Shapiro, B., Shiferaw, B. and Twomlow, S. (2008). Coping better with current climatic variability in the rain-fed farming systems of

---

* Corresponding Author

sub-Saharan Africa: An essential first step in adapting to future climate change? *Agriculture, Ecosystems and Environment* 126: 24–35.

[7]   IPCC (2007). Regional climate projections. In: *Climate Change 2007: The Physical Science Basis. Contribution of Working Group 1 to the Fourth Assessment Report of the Intergovernmental Panel on Climate Change.* (Eds S. Solomon, D. Quin,M.Manning, Z. Chen, M. Marquis, K. B. Averyt, M. Tignor and H. L. Miller). Cambridge, UK: Cambridge University Press.

[8]   IISD (2003). *Livelihoods and Climate Change: Combining disaster risk reduction, natural resource management and climate change adaptation in a new approach to the reduction of rural poverty.* IUCN-IISD-SEI-IC Task Force on Climate Change. Published by IISD.

[9]   Hawkins, E. and Sutton, R. T. (2009). The potential to narrow uncertainty in regional climate predictions./ <http://centaur.reading.ac.uk/1766/> Bulletin of the American Meteorological Society, 90 (8). pp. 1095-1107. ISSN 1520-0477

[10] Hawkins, E. and Sutton, R. T. (2011). The potential to narrow uncertainty in projections of regional precipitation change./ <http://centaur.reading.ac.uk/5716/> Climate Dynamics, 37 (1-2). pp. 407-418. ISSN 1432-0894

[11] Lobell, D, Burke, M, Tebaldi, C, Mastrandrea, M, Falcon, W, Naylor, R. (2008) Prioritizing climate change adaptation needs for food security for 2030. *Science* 319: 607–610

[12] Thornton PK, Jones PG, Owiyo TM, Kruska RL, Herrero M, Kristjanson P, Notenbaert A, Bekele N, Omolo A, with contributions from Orindi V, Otiende B, Ochieng A, Bhadwal S, Anantram K, Nair S, Kumar V, Kulkar U. (2006). *Mapping climate vulnerability and poverty in Africa.* Report to the Department for International Development. International Livestock Research Institute, Nairobi, Kenya. 200pp

[13] Williams, A.P. and Funk, C (2010). A westward extension of the warm pool leads to a westward extension of the Walker circulation, drying eastern Africa. *Climate Dynamics.* DOI 10.1007/s00382-010-0984-y . Published January 2011on line with Open Access by Springerlink.com

[14] Rao, K. P. C., Ndegwa, W. G., Kizito, K. and Oyoo, A. (2011) Climate variability and change: Farmer perceptions and understanding of intra-seasonal variability in rainfall and associated risk in semi-arid Kenya. *Experimental Agriculture* 47: 267–291.

[15] Osbahr, H., Dorward, P. Stern, R. D and Cooper, S. J. (2011) Supporting agricultural innovation in Uganda to climate risk: linking climate change and variability with farmer perceptions. *Experimental Agriculture* 47: 293–316.

[16] IPCC (2001). Watson, R. T.; and the Core Writing Team, ed., Climate Change 2001: Synthesis Report, Contribution of Working Groups I, II, and III to the Third Assessment Report of the Intergovernmental Panel on Climate Change, Cambridge University Press, ISBN 0-521-80770-0 (pb: 0-521-01507-3).

[17] Brown, C., Greene A. M., Block, P. J. Giannini, A. (2008) Review of downscaling methodologies for Africa Applications.IRI Technical Report. pp 31, http://academiccommons.columbia.edu/catalog/ac:126383

[18] Burton, I. and van Aalst, M. (2004). Look before you leap: A risk management approach for incorporating climate change adaptation into World Bank Operations, *World Bank Monograph*, Washington (DC), DEV/GEN/37 E.

[19] DFID (2005). *Climate Proofing Africa: Climate and Africa's development challenge.* Department for International Development, London.

[20] Washington, R., Harrison, M., Conway, D., Black, E., Challinor, A.J., Grimes, D., Jones, R., Morse, A. Kay, G. and Todd, M. (2006) African climate change: taking the shorter route. *Bulletin of the American Meteorological Society* 87:1355–1366.

[21] Cooper, P. J. M, Rao, K. P. C, Singh, P, Dimes, J, Traore, P. S, Rao, K, Dixit, P. and Twomlow, S. J. (2009). Farming with current and future climate risk: Advancing a 'Hypothesis of Hope' for rainfed agriculture in the semi- arid tropics. *Journal of SAT Agricultural Research* 7: 1–19.

[22] Matlon, P. and Kristjanson, P., (1988). Farmer's Strategies to Manage Crop Risk in the West African Semi-arid Tropics. In: Unger, P.W., Jordan, W.R., Sneed, T.V. and Jensen R.W. (Eds), Challenges in Dryland Agriculture: a Global Perspective. Proceedings of the International Conference on Dryland Farming, August 15 – 19, 1988, Bushland Texas, USA pp 604 – 606.

[23] Van de Steeg J, Herrero M, Kinyagi J, Thornton PK, Rao KPC, Stern R, Cooper, P (2009). *The influence of climate variability and climate change on the agricultural sector in East and central Africa – Sensitizing the ASARECA strategic plan to climate change.* Research Report 22, ILRI. pp 85. ISBN 92-9146-238-1 [PDF available from authors, ASARECA and ILRI]

[24] Moore, N., Alagarswamy, G., Pijanowski, B., Thornton, P., Lofgren, B., Olson, J., Andresen, J., Yanda, P. and Qi, J. (2011) East African food security as influenced by future climate change and land use change at local to regional scales *Climatic Change* DOI 10.1007/s10584-011-0116-7. Published on line June 2011 by Springerlink.com

[25] Intergovernmental Panel on Climate Change IPCC website (http://www.ipcc-data.org/) accessed March 2012.

[26] Decision and Policy Analysis Program of the International Centre for Tropical Agriculture (CIAT) http://www.ccafs-climate.org/ accessed March 2012.

[27] Climate Change Agriculture and Food Security Programme (http://ccafs.cgiar.org) accessed March 2012

[28] Bonifacio, R (2008). *A Preliminary Look at Long Range Weather Patterns in Sudan.* Technical Note SIFSIA-N 03/2008. FAO.

[29] Stern, R. D. and Cooper, P. J. M. (2011) Assessing climate risk and climate change using rainfall data–a case study from Zambia. *Experimental Agriculture* 47: 241–266.

[30] Gathenya, M., Mwangi, H., Coe. R and Sang, J. (2011). Climate- and land use-induced risks to watershed servicesin the Nyando River Basin, Kenya. *Experimental Agriculture* 47: 339–356. [PDF available from authors]

[31] Dixit, P. N., Cooper, P. J.M., Rao, K. P. and Dimes, J. (2011) Adding value to field-based agronomic research through climate risk assessment: A case study of maize production in Kitale, Kenya. *Experimental Agriculture* 47: 317–338.

[32] Hansen, J. W., Mason, S. J., Sun, L. and Tall, A. (2011) Review of Seasonal Climate Forecasting for Agriculture in Sub-Saharan Africa. *Experimental Agriculture* 47: 205–240.

[33] University of Reading (2008) Instat+™ - *an interactive statistical package.* Statistical Services Centre, University of Reading, UK

[34] VSN International (2010). *Genstat for Windows.* 13th Edition. VSN International Hemel Hempstead, UK

[35] Frere, M. and Popov, G. (1986) Early agrometeorological crop yield assessment. Food and Agricultural Organization of the United Nations, Rome, Italy. FAO Plant protection Paper 73.

[36] Omumbo, J., Lyon, B., Waweru, S.M., Connor, S.J. and Thomson, M.C. (2011) Raised temperatures over the Kericho tea estates: revisiting the climate in the East African highlands malaria debate. *Malaria Journal* 2011, 10:12 doi:10.1186/1475-2875-10-12

[37] Allan, A. Y. (1972). *The influence of agronomic factors on maize yields in Western Kenya with special reference to time of plantings.* PhD Thesis, University of East Africa, Uganda.

[38] Kirkpatrick D. & Kirkpatrick J. (2006) Evaluating Training Programs: The Four Levels (Third Edition). *San Francisco: Berrett-Koehler Inc.*

[39] Dataverse Network at Harvard University (http://dvn.iq.harvard.edu/dvn/) accessed March 2012.

Chapter 2

Impact of Climate Change on Vegetation and Permafrost in West Siberia Subarctic

Nataliya Moskalenko

Additional information is available at the end of the chapter

1. Introduction

The goal of this ongoing study is to examine the impact of climate change on vegetation and permafrost in ecosystems of West Siberia Subarctic. Results of long-term monitoring of northern taiga ecosystem under impact of climatic changes are presented.

The warming of an observable climate from the end of 20th century was accompanied by changes of vegetation and permafrost degradation, especially in the zone of sporadic permafrost. This important problem is examined in works of many researchers (Tyrtikov, 1969, 1979; Belopukhova, 1973; Brown, Pewe, 1973; Nevecherya et al, 1975; Yevseyev V.P, 1976.; Nelson et al. 1993; Ershov et al. 1994; Pavlov 1997, 2008; Moskalenko,1999; Osterkamp et al. 1999; Parmuzin & Chepurnov 2001; Izrael et al. 2002, Kakunov & Sulimova 2005; Hollister, Webber & Tweedie, 2005; Walker et al. 2006; Perlstein et al. 2006; Oberman 2007; Leibman et al. 2011). They demonstrated that freezing and thawing conditions change in response to the vegetation dynamics. Increases in moss and lichen cover thickness result in the reduction of active layer thickness, and decreases in soil and ground temperatures. However in these works not enough attention was given to estimated impact of climate on the vegetation and permafrost in the ecosystems. In the present report the author tries to fill this deficiency based on long-term monitoring of changes in the northern taiga ecosystem of Western Siberia.

2. Location and parametric considerations

Research on ecosystems were carried out since 1970 on the Nadym stationary site (Fig. 1), located 30 km to a southeast from the town of Nadym (Moskalenko, 2006) in the zone of sporadic permafrost distribution (Melnikov, 1983). Patches of permafrost, occupying up to 50% of areas, are closely associated with peatlands, peat bogs, and frost mounds of III fluvial-lacustrine plain having elevations ranging from 25 to 30m above sea level. The plain is composed of sandy deposits interbedded with clays, with an occasional covering of peat (Andrianov et al. 1973).

During ecosystem monitoring were used remote and cartographical methods. Office studies and field decoding of remote sensing materials from 1970 up to 2009 was added by land route and detail field descriptions on permanent transects and 10x10m plots, fixed on a terrain. Leveling of permanent marks was carried out by electronic level Sprinter 150M every year. Two times for observation period near plots biomass resources were determined. Repeated mapping of vegetation was performed on 1x1m permanent grids for studying of vegetation structure and dynamics. Annual geobotanical descriptions are made on 28 permanent fixed (10 x 10 m) plots. The structure, average height, phenological and vital condition, frequency and coverage of plant species on 50 registered 0.1m² plots were recorded.

Study of spatial and temporal patterns of active layer thickness, caused with microrelief and vegetation mosaic was carried out on 100x100m CALM (Circumpolar Active Layer Monitoring) grid. On 121-grid nodes detail vegetation descriptions and repeated leveling of microrelief were performed. It would reveal some correlations between active layer thickness, vegetation and microrelief. In 16 10-m boreholes and 1 30-m borehole were established loggers Hobo, and measurements of permafrost temperature were carried out by project TSP (Thermal State of Permafrost). Air and soil temperatures were measured too. Monthly average and mean annual temperatures of air and grounds in a wood and on a peatland are resulted in tables 1 and 2.

3. Investigations and observations

Ecosystem changes have been revealed as a result of 40-years observation over a microrelief, species composition of a vegetation cover, height, frequency and coverage of dominant species of plants, soil and permafrost temperature, thickness and moisture of active layer on permanent plots and transects.

3.1. Impact of increase in amount of atmospheric precipitation on vegetation and permafrost

The analysis of the received data has allowed to revealing tendencies in development of a natural vegetation cover. In wood communities in connection with increase of atmospheric precipitation amount which is marked last decades, the increase in participation of mosses, and change of green moss-lichen sparse forests by lichen-green moss plant communities on drained sites is observed. Changes of atmospheric precipitation (Fig. 2) and *Cladina rangiferina* frequency (Fig. 3) in Birch-pine sparse forest are presented. Coverage of *Pleurozium Schreberi* opposite increases (Fig. 4).

In connection with the increase of atmospheric precipitation process of bog formation on flat poorly drained surfaces of plains becomes more active. As a result hummocky pine cloudberry-wild rosemary-lichen-peat moss open woodlands were replaced by *andromeda*-cotton grass-sedge-peat moss bogs. Hummocks settled, and the lenses of permafrost under hummocks thawed.

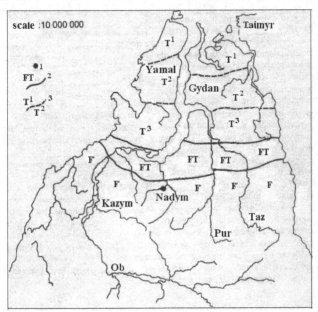

1– site, 2 – boundaries of zones (T – tundra, FT – forest tundra, F – taiga) 3 – boundaries of tundra subzones (T^1 – northern, T^2 – typical, T^3 – southern).

Figure 1. Location of the Nadym site

Depth, m, year		Months												Year
		1	2	3	4	5	6	7	8	9	10	11	12	
Air	a	-17,8	-18,5	-14,6	-9,5	2	8,7	15,9	11,6	5,9	-3,1	-14	-16,1	-4,1
	b	-24,3	-28,2	-14,5	-6,4	-2,1	10,1	15,1	11,4	8	-2,4	-21,4	-33,8	-7,4
0	a	-2,7	-2,5	-3,1	-2,2	-0,16	7	12,9	11,5	5,8	0,49	-2,4	-1,8	1,9
	b	-1,8	-2,5	-2,2	-1,2	-0,1	5,1	11,8	9,8	7,6	1	-2,6	-3,1	1,3
0,25	a	-0,3	-0,5	-0,9	-1,0	-0,3	0,0	5,5	8,1	6	2,3	0,5	0,1	1,6
	b	0	-0,2	-0,5	-0,5	-0,1	-0,1	5,7	7,8	6,6	3,2	0,4	0	1,8
0,5	a	0,2	0	-0,4	-0,6	-0,1	0,0	3,6	6,8	5,9	2,9	1,2	0,6	1,7
	b	0,4	0,2	0	-0,2	0	0	3,8	6,4	6,2	3,8	1,2	0,5	1,9
1	a	0,5	0,3	0,1	-0,1	-0,1	0,0	2,0	5,3	5,3	3,3	1,7	1	1,6
	b	0,7	0,5	0,2	0,1	0,1	0,2	2,4	4,8	5,6	4,1	1,9	1	2,1
1,5	a	0,8	0,8	0,4	0,2	0,2	0,2	1,3	4,3	4,7	3,5	2	1,3	1,6
	b	1	0,7	0,5	0,3	0,3	0,3	1,5	3,9	4,8	4	2,2	1,3	2,2
3	a	2,1	1,6	1,4	1	1	0,9	0,8	1,5	2,0	2,0	1,5	1,0	1,4

Table 1. Monthly average and mean annual temperatures of air and grounds in the wood (a -2008, b - 2009).

Depth, m, year		Months												Year
		1	2	3	4	5	6	7	8	9	10	11	12	
Air	a	-17,2	-18,9	-15,7	-11	0,8	8,5	15,8	11,7	6,1	-2,8	-13,8	-16	-4,4
	b	-24,2	-28,3	-15,4	-8,2	-3,9	9,5	15	13,8	8,3	-2,1	-21,2	-33,3	-7,5
0	a	-2,5	-2,5	-2	-0,9	1,1	7,9	13,3	11,3	5,5	0,12	-1,6	-0,9	2,1
	b	-2,5	-2,5	-2	-0,9	1,1	7,9	13,3	11,3	5,5	0,12	-1,6	-0,9	2,1
0,25	a	-0,5	-0,6	-0,7	-0,5	-0,2	0,8	3,8	5,6	3,9	0,5	-0,1	-0,1	1,0
	b	-0,3	-0,8	-0,8	-0,5	-0,2	0,7	4,2	6,3	4,8	1,6	-0,1	-0,3	1,2
0,5	a	-0,1	-0,1	-0,3	-0,3	-0,1	-0,1	1,0	3,9	3,1	0,5	-0,0	-0,0	0,6
	b	-0,1	-0,1	-0,3	-0,3	-0,1	-0,1	0,8	4	3,8	1,5	0	0	0,8
1	a	-0,0	-0,0	-0,0	-0,0	-0,0	-0,0	-0,0	1,9	1,8	0,3	-0,0	-0.0	0,3
	b	-0,0	-0,0	-0,00	-0.1	-0,0	-0,0	-0,0	1.1	1,6	1,5	0,8	0,2	0
1,5	a	-0,1	-0,1	-0,1	-0,1	-0,1	-0,0	-0,0	0,6	0,7	0,1	-0,1	-0,1	0,1
	b	-0,1	-0,1	-0,1	-0,1	-0,1	-0,1	0	0,2	0,6	0,5	0,3	0	0
3	a	-0,1	-0,1	-0,1	-0,1	-0,1	-0,1	-0,1	-0,1	-0,1	-0,1	-0,1	-0,1	-0,1
	b	-0,1	-0,1	-0,1	-0,1	-0,1	-0,1	-0,1	-0,1	-0,1	-0,1	-0,1	-0,1	-0,1

Table 2. Monthly average and mean annual temperatures of air and grounds on the peatland (a -2008, b -2009).

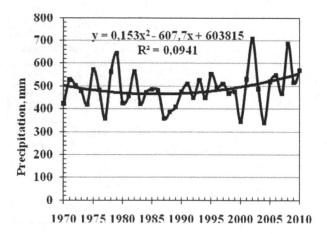

Figure 2. Amount of atmospheric precipitation

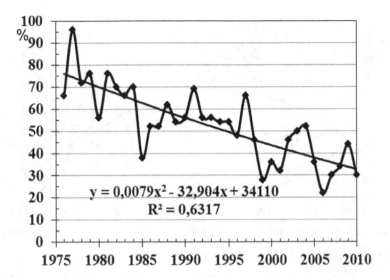

Figure 3. Frequency changes of *Cladina rangiferina* in Birch-pine sparse forest

The frequency of wild rosemary *(Ledum palustre)* which dominated in a cover of the open woodland fell sharply after 1997 (Fig. 5, 2). The frequency of cotton-grass *(Eriophorum angustifolium)* for the past decade increased, and it began to dominate the cover (Fig. 5, 1).

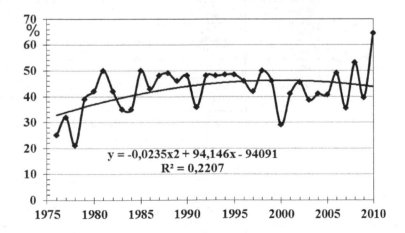

Figure 4. Coverage changes of *Pleurozium schreberi* in Birch-pine sparse forest

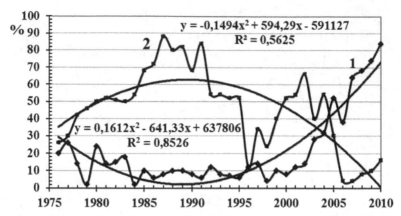

Figure 5. Frequency changes of *Ledum palustre* (2) and *Eriophorum angustifolium* (1) on flat boggy site

Comparison of biomass in wood communities and bog communities shows that by bog formation in wood all aboveground biomass decreases from 2316 to 1715 g/m² and biomass of graminoid and mosses increases (table 3). Comparison of species composition of wood and bog plant communities presents that biodiversity of vegetation cover in process of bogginess decreases in the result of absence mesophyte species of sedges and shrubs (*Carex globularis, Empetrum nigrum, Vaccinium vitis-idaea*), and also lichens (*Cladina rangiferina, C. stellaris, Cetraria islandica, Cladonia coccifera*, table 4). Common number of species decreases from 27 to 17.

Vegetation		Wood	Bog	Tundra
Deciduous shrubs	Stems	41	84	10
	Live leaves	9	23	1
	Dead leaves	1	0	0
	Berries	0,5	1	0
Evergreen shrubs	Stems	141	141	141
	Live leaves	66	84	33
	Dead leaves	2	4	1
	Berries	0,2	2	1
Graminoid	Live leaves	0.3	3	14
	Dead leaves	0.3	19	46
Forb		2	13	3
Mosses	Live	80	383	1
	Dead	274	272	1
Lichens	Live	812	228	930
	Dead	400	104	524
Litter		490	317	215
All biomass		2316	1715	1926

Table 3. Aboveground biomass (g/m²) of different plant communities on the Nadym site.

Species	Year	Height, cm	Coverage, %	Frequency, %
1. Andromeda polifolia	1	7	2	54
	2	12	4	72
	3	15	5	76
	4	15	3.5	72
	5	15	1	54
2. Betula nana	1	45	2	16
	2	65	1	30
	3	80	2.5	32
	4	80	0.8	14
	5	80	0,1	2
3. Calamagrostis lapponica	1	30	<1	2
	2	70	0.1	12
	3	25	0.5	10
	4	60	<1	<1
	5	50	<1	<1
4. Carex globularis	1	20	4	32
	2	25	7	64
	3	30	4.5	52
	4	35	0.1	2
	5	-	-	-
5. Carex rotundata	1	20	<1	10
	2	30	1	12
	3	60	0.5	4
	4	50	0.1	4
	5	30	1.5	28
6. Empetrum nigrum	1	4	1	6
	2	10	1	10
	3	10	1	14
	4	8	0.1	2
	5	-	-	-
7. Eriophorum angustifolium	1	30	4	20
	2	50	0.2	10
	3	75	1.5	6
	4	100	2	52
	5	60	10.5	84
8. Eriophorum vaginatum	1	30	1	1
	2	60	0.2	10
	3	30	1	14
	4	50	0.1	2
	5	60	3	14
9. Juncus filiformis	1	15	0.1	4
	2	35	0.1	4
	3	40	1	8
	4	30	2	8
	5	40	2	6

Species	Year	Height, cm	Coverage, %	Frequency, %
10. Ledum palustre	1	30	7	32
	2	40	3	48
	3	40	8	60
	4	40	1.5	2
	5	40	1.5	2
11. Oxyccocus microcarpus	1	1	5	44
	2	2	2	30
	3	2	1.5	30
	4	2	0.1	2
	5	2	1	10
12. Pinus silvestris	1	300	<1	<1
	2-4	-	-	-
	5	45	<1	<1
13. Rubus chamaemorus	1	3	14	52
	2	9	3	46
	3	10	8	52
	4-5	-	-	-
14. Vaccinium myrtillus	1	3	1	28
	2	10	3	44
	3	15	1.5	42
	4-5	-	-	-
15. Vaccinium uliginosum	1	20	6-	54
	2	30	10	60
	3	30	15	72
	4	40	0.1	2
	5	40	0.6	16
16. Vaccinium vitis-idaea	1	4	<1	<1
	2	6	1	12
	3	10	1	22
	4-5	-	-	-
17. Cetraria islandica	1	1	0.1	2
	2	2	0.2	8
	3	5	1	6
	4-5	-	-	-
18. Cladonia coccifera	1	1	1	4
	2	1	1	4
	3	3	0.1	4
	4-5	-	-	-
19. Cladina rangiferina	1	6	1	4
	2	7	1	22
	3	7	3.5	26
	4-5	-	-	-
20. Cladina stellaris	1	7	4	10
	2	8	3	22
	3	8	0.2	8

Species	Year	Height, cm	Coverage, %	Frequency, %
	4-5	-	-	-
21. Aulacomnium palustre	1	2	0.1	2
	2	2	0.1	2
	3	2	<1	<1
	4	-	-	-
	5	2	<1	<1
22. Dicranum congestum	1	1	0.1	2
	2	1.5	0.1	2
	3	2	3	6
	4	-	-	-
	5	0.5	0.1	4
23. Pleurozium schreberi	1	1	8	20
	2	2	28	42
	3	4	20.5	40
	4	4	0.1	2
	5	4	2	8
24. Polytrichum commune	1	3	3	38
	2	8	16	60
	3	8	21	70
	4	8	0.2	2
	5	8	25.5	66
25. Sphagnum angustifolium	1	1	11	18
	2	4	7	14
	3	4	6	8
	4	4	0.1	2
	5	5	19	48
26. Sphagnum fuscum	1	2	36	52
	2	2.5	21	24
	3	3	25	28
	4	3	0.1	2
	5	3	9.5	20
27. Sphagnum lindbergii	1	4	23	34
	2	8	8	14
	3	8	5	10
	4	8	<1	<1
	5	8	26.5	36

Table 4. Species composition of vegetation on flat boggy site in 1975 (1), 1985 (2), 1995 (3), 2005 (4) and 2010 (5) years.

3.2. Impact of increase in air temperature on vegetation and permafrost

Last decades in the north of Western Siberia rise in air temperature is observed (Fig. 6). Increase of the air thawing index (the sum monthly mean air temperatures above 0°C) caused the appearance on flat and palsa peatlands separate trees (*Betula tortuosa, Pinus sibirica, Pinua silvestris*); increase in frequency and height of shrubs (*Betula nana, Ledum palustre,* Fig. 7) and coverage them of a soil surface. These plant species can serve as indicators of climate warming.

Long-term studying of plants communities and active layer thickness in northern taiga has allowed calculating of plant communities frequency with active layer thickness. The smallest values of active layer thickness (67.1 cm) are observed under *Rubus chamaemorus-Ledum palustre-Sphagnum-Cladina rangiferina* cover on flat peatland (coefficient of correlation -0.71). Areas with deepest active layer thickness (173.7 cm) are confined to large sedge-moss pools within peatlands (coefficient of correlation 0.58).

The analysis of the given measurements of the active layer thickness on palsa peatland (Fig. 8) has shown that it has a trend to the increase, caused by increase in the thawing index of air temperature, which trend for 1970-2010 makes 0.2ºC in a year. The permafrost temperature at the depth of 10m has increased on 1.4ºC. Temperature of permafrost at the depth of 10m (layer with minimum annual fluctuations of temperatures) for the period of researches on the palsa peatland has increased from -1.8ºC up to -0.4ºC (Fig.9, 2). On flat peatland increase of permafrost temperature was less; here permafrost temperature at the depth of 10m has increased from -0.9ºC up to -0.2ºC (Fig.9, 1).

Increase in air temperature and rise in amount of atmospheric precipitation promoted faster recovery of a vegetation cover after a fire. For example, on frost mounds with *Pinus sibirica-*wild rosemary-peat moss-lichen open woodland in 35 years after the fire *Betula nana-*wild rosemary-peat moss-lichen community with *Pinus sibirica* in height 2m had developed (Fig.10).

On the permanent plot located on a flat southern slope the frost mound in height of 3 m. In a well-defined microrelief of tussocks and hummocks height up to 0.8m are characteristic. Pools were usual, sometimes filled with water.

Soil is sandy peat-gley, and frozen at 0.5m depth. Average peat horizon thickness is 30cm. A crown density of *Pinus sibirica* is 0.1, its height 7-8m. The coverage of grasses and dwarf shrubs makes up 40-50%.

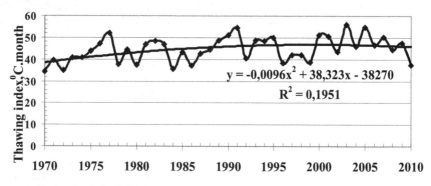

$$y = -0,0096x^2 + 38,323x - 38270$$
$$R^2 = 0,1951$$

Figure 6. Air thawing index in Nadym

The grass-dwarf shrub cover has two-layer structure: the upper layer in height is 0.3-0.35m composed of wild rosemary and *Betula nana,* and the lower layer in height is 0.05-0.15m with abundant cowberry (*Vaccinium vitis-idaea*), *Chamaedaphne calyculata*, cloudberry and sedge (*Carex globularis*). Peat mosses and lichens make up the continuous ground cover.

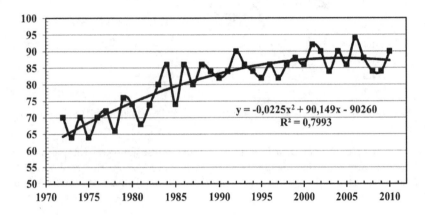

Figure 7. Frequency of *Ledum palustre* on the flat peatland

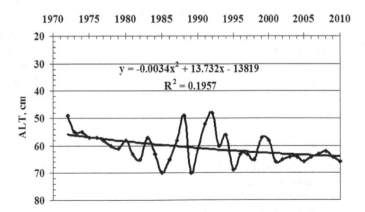

Figure 8. Active layer thickness on the palsa peatland

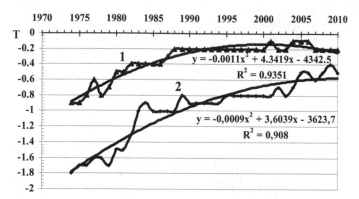

Figure 9. Permafrost temperature (T °C) at the depth of 10m on the flat peatland (1) and the palsa peatland (2)

In June 1976 the plot of grass-dwarf shrub cover, and a forest stand was completely burned. Within two months following the fire the surface cover of 25% consisted of shoots of *Carex globularis*, *Betula nana*, wild rosemary, and cloudberry. In pools the moss coverage of up to 30% was maintained.

One year following the fire the sedge-cloudberry-peat moss grouping was formed, and the next year it was replaced by cloudberry-sedge-wild rosemary-peat moss community. This was the result of the fast recovery of a former role of wild rosemary (Fig.11, 1). In this community the coverage of grasses and dwarf shrubs increased up to 35%, and mosses up to 40%. The next years the coverage of grasses and dwarf shrubs reached its initial value (40-50%), but mosses still covered less than half of plot surface. The frequency of *Betula nana* has increased in 3 times, probably, in connection with the rise in air temperature (Fig. 11, 2).

The occurrence of lichens sharply decreased after the fire, and within 16 years had considerably increased. Only the frequency of *Cladonia coccifera* strongly increased after the fire. The frequency of *Cladina stellaris* was recovered (Fig. 12, 1), and its reduction last decade is connected to increase of amount of atmospheric precipitation that is observed and in undisturbed conditions. The frequency of *Sphagnum fuscum* (Fig. 12, 2) while remains in 2 times less than in initial community. The increase in height of shrubs (*Betula nana*, *Chamaedaphne calyculata*, *Ledum palustre*) is marked also. Changes in species composition, height, coverage and frequency of plants on frost mound are presented in the table 5.

On the cloudberry-wild rosemary-lichen palsa peatlands n 40 years after the fire the cloudberry- *Betula nana*- wild rosemary-lichen-*Polytrichum* communities are found. These communities differ from the initial communities by ground vegetation composition (smaller percentage of lichens) and increase in presence of *Betula nana*. The last, apparently, is connection to the increase of the air thawing index and a snow thickness over the last decades.

In 1971, plot on palsa peatland on which in 1970 were carried out the detailed description of a vegetation cover, measurements of active layer thickness and permafrost temperature, was burned.

This plot is located at top of peat hillocky with height of 2m and with cloudberry-wild rosemary-lichen plant community. In the microrelief of plot are characteristic small *Dicranum* hummocks with heights of 0.1-0.3m and pools with bog dwarf shrubs (*Andromeda polifolia, Chamaedaphne calyculata*) and mosses. The soil of the plot is peaty, and maximum thickness of the active layer is 0.6m. The coverage of grasses and dwarf shrubs equaled 45 %; mosses and lichens - 90%. In a grass-shrub cover two layers are found: an upper layer in height of 0.2-0.4m made up of wild rosemary and *Betula nana*, and a lower layer in height up to 0.15m formed of cloudberry and cowberry. In ground vegetation, lichens preedominated over a *Cladina* genus and frequent *Dicranum* mosses but with low coverage.

In 1975, four years after the fire at the top of the peaty hillocky where the vegetation had been described in 1970, a permanent 10 x 10m plot on the soil surface was established. On this plot, since 1975 on present time, annual geobotanical descriptions are performed.

A 10-meter borehole was drilled at the hillocky top near to the geobotanical plot. According to the drilling the peat thickness is 1m, below lies sand with layers of the clay, underlaying with depth 3,75m by clay. From 1975 year-round temperature measurements of soil and permafrost were observed (Fig. 13). Since 2001 year-round measurements of temperature by loggers are obtained. Thickness and moisture of the active layer were measured.

In four years since the fire on hillocky the cotton-grass-cloudberry-*Polytrichum* community is found in which the coverage of grasses made 15%, and mosses 50 %. After the fire the number of species on the plot was 42% of their common number in 1970. Change of species number could be still large, but appearance of new grass species (*Erophorum russeolum, Carex limosa, Chamaenerium angustifolium*) and shoots of a birch (*Betula tortuosa*) compensated for significant decrease of species number. It has been related to disappearance of five dwarf shrubs (*Vaccinium uliginosum, V. vitis-idaea, Empetrum nigrum, Andromeda polifolia* and *Chamaedaphne calyculata*), *Eriophorum vaginatum*, one species of lichens (*Alectoria ochroleuca*) and three species of mosses (*Sphagnum fuscum, Pleurozium schreberi, Hylocomium splendens*). In the first years of vegetation recovery the frequency and coverage of *Polytrichum* mosses strongly increased (Table 6). Occurrence of dwarf shrubs has decreased, bog grasses have appeared absent earlier, and the occurrence of shrubs increased.

In five years after the fire on hillocky landscape with cotton-grass-cloudberry-*Polytrichum* community the coverage of grasses was 20%, and mosses 50%. The next year there was an appreciable increase in occurrence of *Betula nana* that led to changes of the grass-moss community with *Betula nana*- cloudberry-cotton-grass-*Polytrichum* community.The coverage of grasses and dwarf shrubs in this community gradually grew and in 14 years after the fire had reached its initial value. At this time an appreciable role of wild rosemary began to occur. The ground vegetation by this time covered up to 85% of a plot surface, but it still has consisted of *Polytrichum* mosses. The thickness of the active layer in this plant community has increased up to 65-70cm.

The frequency of lichens though has increased, but the coverage on the surface did not exceed 1-3 %. However the coverage of lichens gradually continued to increase, and in 23 years after the fire it has reached 8.5 %. The coverage of lichens has increased for 40th year up to 18.5%, and includes *Betula nana*-wild rosemary-cloudberry-*Cladina*- *Polytrichum* community in which the occurrence of cotton-grass has decreased. The number of dwarf shrubs and mosses by this time has appreciably increased, but remained less than in undisturbed cover due to the absence of bog dwarf shrubs (*Andromeda polifolia, Chamaedaphne calyculata*) and one species of mosses (*Hylocomium splendens*). The bog grasses which have appeared at early stages of plant community recovery in 2005 have disappeared from the plant community.

A

B

Figure 10. Frost mound before fire (A) and 35 years after it (B)

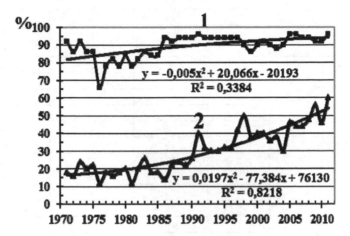

Figure 11. Frequency of Ledum palustre (1) and Betula nana (2) on the frost mound

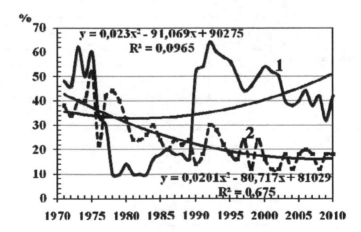

Figure 12. Frequency changes of *Cladina stellaris* (1) and *Sphagnum fuscum* (2) on the frost mound

Species	Year	Height, cm	Coverage, %	Frequency, %
1. Andromeda polifolia	1	10	1	18
	2	13	0.1	8
	3	15	1	14
	4	15	0.1	6
	5	15	0.2	12
2. Betula nana	1	45	2	22
	2	65	1	18
	3	65	1.5	22
	4	80	7	46
	5	100	6	46
3. Carex globularis	1	15	6	64
	2	35	15	80
	3	30	16	86
	4	40	4	96
	5	35	2	84
4. Chamaedaphne calyculata	1	15	4	56
	2	30	1	24
	3	30	7	36
	4	40	2.5	62
	5	40	1	54
5. Empetrum nigrum	1	7	0.1	6
	2	10	0.2	10
	3			
	4	10	0.2	16
	5	10	0.2	10
6. Eriophorum vaginatum	1	10	<1	<1
	2	10	0.4	2
	3	20	0.1	2
	4	20	0.1	2
	5	30	<1	<1
7. Ledum palustre	1	40	15	86
	2	50	9	84
	3	50	20	94
	4	55	21.5	96
	5	55	30	92
8. Oxyccocus microcarpus	1	1	3	46
	2	2	3	30
	3	1	3	30
	4	2	0.9	18
	5	2	0.2	20

Species	Year	Height, cm	Coverage, %	Frequency, %
9. Pinus sibirica	1	800	<1	<1
	2	35	0.1	4
	3	60	<1	<1
	4	170	0.1	2
	5	200	0.1	4
10. Rubus chamaemorus	1	5	5	72
	2	10	11	84
	3	10	6.5	68
	4	12	3	66
	5	10	1.5	46
11. Vaccinium myrtillus	1	10	0.1	2
	2	10	<1	<1
	3	10	0.1	2
	4	12	0.1	4
	5	12	<1	<1
12. Vaccinium uliginosum	1	17	0.1	4
	2	25	0.1	2
	3	25	1	2
	4	25	0.2	8
	5	25	0.1	2
13. Vaccinium vitis-idaea	1	7	5	82
	2	10	11	88
	3	15	4	86
	4	15	6.5	86
	5	20	7	84
14. Cetraria cucullata	1	4	0.2	10
	2	4	<1	<1
	3	4	0.1	2
	4	5	0.1	4
	5	5	0.4	2
15. Cetraria islandica	1	4	0.2	10
	2	4	0.1	6
	3	4	15	16
	4	5	0.1	6
	5	5	0.2	8
16. Cladonia amaurocraea	1	3	0.1	2
	2	3	0.1	2
	3	4	0.2	10
	4	5	1.5	12
	5	8	0.8	8
17. Cladonia coccifera	1	3	0.1	2
	2	2	1	54
	3	4	10	52
	4	5	2.5	32
	5	7	2.5	22

Species	Year	Height, cm	Coverage, %	Frequency, %
18. Cladina rangiferina	1	8	19	60
	2	5	0.5	26
	3	5	0.7	36
	4	9	5.5	32
	5	9	2	32
19. Cladina stellaris	1	8	27	60
	2	4	0.4	18
	3	4	1	56
	4	9	7.5	48
	5	10	12	42
20. Dicranum congestum	1	1	0.1	2
	2	1	2	2
	3	2	0.5	4
	4	2	<1	<1
	5	2	<1	<1
21. Pleurozium schreberi	1	2	2	52
	2	2	0.1	2
	3	3	3	8
	4	3	0.4	2
	5	3	3	12
22. Polytrichum commune	1	5	0.1	4
	2	3	2	24
	3	3	7	20
	4	3	0.2	6
	5	3	3	16
23. Sphagnum angustifolium	1	2	0.1	2
	2	3	14	22
	3	3	8	26
	4	3	10	18
	5	3	6	12
24. Sphagnum fuscum	1	2	23	52
	2	3	14	14
	3	3	6	8
	4	3	16.5	18
	5	3	14.5	18
25. Tomenthypnum nitens	1	1	2	2
	2	1	0.1	2
	3	1	0.8	4
	4	1	0.1	2
	5	2	<1	<1

Table 5. Species composition of vegetation on the frost mound in 1975 (1), 1985 (2), 1995 3), 2005 (4) and 2010 (5) years.

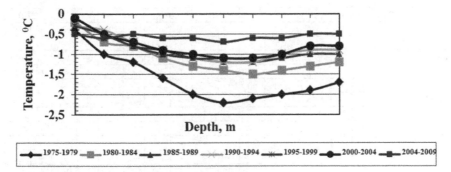

Figure 13. Five-year moving averages of ground temperatures at the depths of 1-10 m on the palsa peatland

Negatively reacted to a fire some shrubs (*Vaccinium vitis-idaea, Ledum palustre*), lichens, green mosses and *Sphagnum fuscum*. In 15 years after a fire at the cowberry, the wild rosemary and all before plentiful species of lichens (*Cladina, Cetraria*), the frequency and the coverage strongly differed from initial sizes. This distinction was kept and in 23 years after fire. Participation of some species of lichens (*Cladonia coccifera* and *Cladonia amaurocraea)* and blueberries for the investigated period was recovered. At cloudberries sizes of the coverage were made even to initial sizes, but it frequency still was more than in 2 times smaller.

In 30 years after the fire the frequency of *Betula nana* has exceeded initial size, the frequency of *Ledum palustre* too has considerably increased and for 40-th year was only a little less, than in not disturbed community. Only at the cowberry and dominant species of lichens (*Cladina stellaris* and *Cladina rangiferina*) the frequency for all period of observations was not recovered. The analysis of frequency diagrams of *Betula nana* and *Ledum palustre* (Fig. 14) shows, that there is a positive trend which will be coordinated to increase of summer air temperatures.

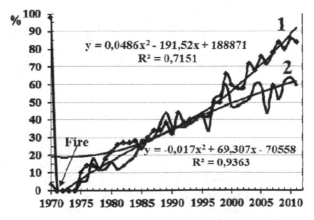

Figure 14. Frequency of *Ledum palustre* (1) and *Betula nana* (2) on the palsa peatland

Species	1970		1975		1980		1985		1990		1995		2000		2005		2010	
	1	2	1	2	1	2	1	2	1	2	1	2	1	2	1	2	1	2
Andpol	2	7	-	-	-	-	-	-	-	-	-	-	-	-	-	-	1	10
Betnan	4	40	5	35	16	50	28	70	36	70	44	70	48	80	60	90	64	100
Bettor	-	-	1	45	1	130	1	200	1	300	1	400	1	500	1	600	4	600
Callap	4	30	-	=	1	30	1	60	1	20	-	-	-	-	-	-	1	40
Carglo	2	25	-	-	-	-	-	-	1	20	1	20	2	35	1	30	1	35
Carlim	-	-	2	20	2	20	-	-	-	-	-	-	-	-	-	-	-	-
Carrot	-	-	-	-	2	20	-	-	-	-	-	-	-	-	-	-	-	-
Chaang	-	-	2	30	1	40	1	35	-	-	-	-	-	-	-	-	-	-
Chacal	2	17	-	-	-	-	-	-	-	-	-	-	-	-	-	-	-	-
Empnig	2	10	-	-	4	10	-	-	1	10	1	10	2	10	1	10	1	10
Eriang	-	-	-	-	2	20	1	30		20	1	20	1	10	-	-	-	-
Erirus	-	-	45	35	46	30	64	30	82	30	34	15	-	-	-	-	-	-
Erisch	-	-	-	-	12	20	2	30	-	-	-	-	-	-	-	-	-	-
Erivag	10	12	-	-	2	35	16	30	10	50	54	20	68	30	58	50	38	35
Ledpal	98	20	10	15	22	25	24	30	32	35	42	35	60	40	76	45	86	45
Pinsib	-	-	-	-	1	5	-	-	4	5	4	15	1	20	6	35	10	55
Pinsil	-	-	-	-	-	-	-	-	-	-	-	-	2	6	-	-	1	50
Rubcha	98	10	28	5	22	10	34	12	30	12	34	10	32	15	44	15	42	15
Vaculi	1	10	-	-	2	10	1	20	1	20	1	30	1	30	1	35	1	35
Vacvit	46	5	-	-	6	7	6	7	2	7	4	7	4	7	1	10	1	10
Aulpal	6	2	-	-	-	-	-	-	-	-	-	-	-	-	-	-	-	-
Diccon	10	2	12	1	20	1	1	1	1	2	1	4	2	1	2	1	1	2
Hylspl	2	1	-	-	-	-	-	-	-	-	-	-	-	-	-	-	-	-
Plesch	4	2	-	-	-	-	-	-	-	-	-	-	-	-	-	-	4	2
Polcom	6	2	96	2	96	5	98	6	98	7	96	7	96	7	94	7	98	7
Sphfus	5	2	-	-	-	-	-	-	-	-	-	-	-	-	-	-	4	3
Aleoch	4	3	-	-	-	-	-	-	-	-	2	2	2	3	6	4	6	5
Cetcuc	28	3	6	1	2	1	1	2	2	3	4	3	8	4	8	5	10	5
Cetisl	6	3	4	1	2	2	2	3	2	3	2	4	1	5	2	5	12	6
Cetniv	94	2	10	1	6	2	1	2	1	2	6	3	2	3	8	4	8	4
Claama	14	3	2	1	1	1	4	2	4	3	26	4	18	4	18	5	22	6
Clacoc	20	3	2	1	12	1	36	2	38	3	48	4	40	4	48	5	40	5
Claran	78	8	2	1	8	2	6	3	14	5	34	6	12	6	20	6	30	6
Claste	98	10	10	1	12	2	22	3	38	4	42	5	30	5	38	6	40	6

Table 6. Frequency (1, %) and height (2, cm) changes of plant species on the palsa peatland in 1970-2010 years.

Plant species. *Vascular plants: Andpol – Andromeda polifolia, Betnan – Betula nana, Bettor – Betula tortuosa, Callap – Calamagrostis lapponica, Carglo – Carex globularis, Carlim – Carex limosa, Carrot – Carex rotundata, Chaang – Chamaenerium angustifolium, Empnig – Empetrum nigrum, Eriang – Eriophorum angustifolium, Erirus – Eriophorum russeolum, Erisch – Eriophorum scheucheri, Erivag – Eriohorum vaginatum, Ledpal – Ledum palustre, Pinsib – Pinus sibirica, Pinsil – Pinus silvestris, Rubcha – Rubus chamaemorus, Vaculi – Vaccinium uliginosum, Vacvit – Vaccinium vitis-idaea.*

Mosses: *Aulpal – Aulacomnium palustre, Diccon – Dicranum congestum, Hylspl – Hylocomium splendens, Plesch – Pleurozium schreberi, Polcom - Polytrichum commune, Sphfus – Sphagnum fuscum.*

Lichens: *Aleoch – Alectoria ochroleuca, Cetcuc – Cetraria cucullata, Cetisl – Cetraria islandica, Cetniv – Cetraria nivalis, Claama – Cladonia amaurocraea, Clacoc – Cladonia coccifera, Claran – Cladina rangiferina, Claste – Cladina stellaris.*

Stages of vegetation recovery after the fire on the frost mound and the palsa peatland are presented in Table 7. Comparison of rates of vegetation cover restoration in these ecosystems demonstrate that on flat weakly drained top of frost mound the vegetation recovery is faster than on better drained palsa peatland. The domination in ground vegetation of *Polytrichum* mosses and the lower occurrence of lichens persists longer.

Stages and their duration (years)	Ecosystems	
	I	II
Grass-moss (1-5)	1a	1б
Shrub-grass-moss (6-15)	2a	2б
Shrub-grass-lichen-moss (16-35)	3a	3б
Grass-shrub-moss-lichen (36-50)	4a	4б

Table 7. Stages of vegetation recovery after the fire in different ecosystems

Ecosystems: I – cloudberry-wild rosemary-lichen palsa peatland, II – frost mound with *Pinus sibirica* wild rosemary- peat moss-*Cladina* open woodland.

Plant communities: 1a – cotton grass-cloudberry-Polytrichum, 1б–sedge-cloudberry-peat moss, 2a – *Betula nana*-cloudberry-cotton- grass-*Polytrichum*, 2б – cloudberry-sedge-wild rosemary-peat moss, 3a – cloudberry-*Betula nana*--wild rosemary-*Cladina-Polytrichum*, 3б – *Betula nana*-wild rosemary-peat moss-*Cladina*, 4a - cloudberry-*Betula nana*-wild rosemary-*Cladina-Polytrichum*, 4б – *Pinus sibirica- Betula nana*-wild rosemary-peat moss-*Cladina*.

3.3. Impact of vegetation dynamics on permafrost

On the dwarf shrub-cotton grass-peat moss bogs in the result of vegetation dynamics it is possible to observe formation of new frost heavy hummocks (Fig.16). The height of one of young frost mound, which beginning of formation concerns to 1973, makes by the present moment 80 cm.

The ecosystems are detected, in which the local temperature decrease observed on a background of the general tendency of temperature increase, caused by dynamics of a vegetation cover. It is necessary to allow a possibility of such different tendencies of temperature changes in ecosystems at for the same changes of a climate at geocrylogical monitoring.

For example, such downturn of permafrost temperatures was observed on dwarf shrub-sedge-peat moss bog, replaced through 25 years by sedge-dwarf shrub- lichen-peat moss peatland as a result of increase in moss thickness, accumulation of peat and growths of dwarf shrubs (*Andromeda polifolia, Chamaedaphne calyculata*). Here permafrost temperatures for the investigated period have gone down on 0.3⁰C (Fig.15) though in the next flat peatlands surrounding a drained up bog, the permafrost temperature became higher.

On cotton grass-peat moss bogs with the lowered permafrost table on formed on it dwarf shrub-peat moss hummocks after cold winters it is observed formation of new frozen ground. Mean active layer thickness on these hummocks is 80 cm.

Figure 15. New frost heavy hummocks on the dwarf shrub-cotton grass-peat moss bog

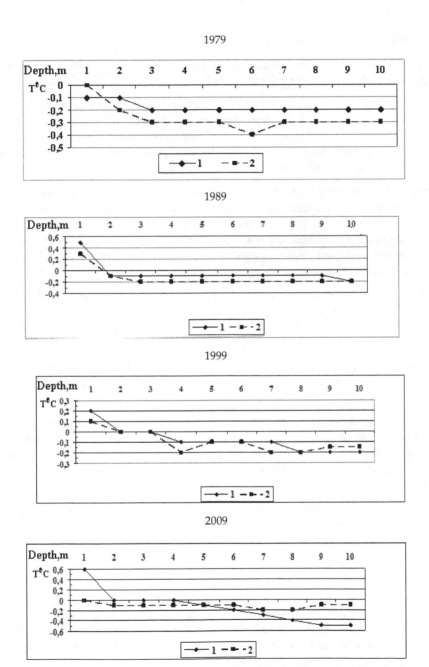

Figure 16. Permafrost temperature (T⁰C) changes on the bog (1) and on the peatland (2) at the depths of 1-10 m in 1979, 1989, 1999 and 2009 years.

4. Results and discussions

Long-term monitoring of the northern taiga ecosystem changes has allowed revealing impact of climatic changes on a vegetation cover and permafrost.

During the last decades in the north of West Siberia the rise in air temperature and the increase in amount of atmospheric precipitation are observed. In wood communities in connection with increase of atmospheric precipitation amount which is marked last decades, the increase in participation of mosses, and change of green moss-lichen sparse forests by lichen-green moss plant communities on drained sites is marked.

On flat poorly drained surfaces of plains process of bog development became more active. As a result of it hummocky pine cloudberry-wild rosemary-lichen-peat moss open woodlands with lenses of permafrost under the hummocks are replaced by andromeda-cotton grass-sedge-peat moss thawed bogs. Comparison wood communities and bog communities show that by bog formation in wood all aboveground biomass decreases on 26% and biodiversity in process of bogginess decreases on 37%.

Increase of the thawing index of air temperature caused the appearance on the flat and palsa peatlands separate trees (*Betula tortuosa, Pinus sibirica, Pinus silvestris*), increase in the frequency and the height of shrubs (*Betula nana, Ledum palustre*) and in the coverage them of a soil surface. These plant species can serve as indicators of climate warming.

The analysis of the given measurements of the active layer thickness on palsa peatland has shown that it has a trend to the increase, caused by increase in the air thawing index, which trend for 1970-2010 makes 0.2ºC in a year.

The permafrost temperature at the depth of 10 m has increased on 1.4ºC. Temperature of permafrost at the depth of 10 m for the period of research on the palsa peatland has increased from -1.8ºC up to -0.4ºC. On the flat peatland increase of the permafrost temperature was less; here the permafrost temperature at the depth of 10 m has increased from -0.9ºC up to -0.2ºC.

In conditions of climate warming fires began to be observed more often. On cloudberry-wild rosemary-lichen palsa peatlands 40 years after a fire are formed cloudberry-Betula nana-wild rosemary-lichen-Polytrichim plant communities. These plant communities differ from initial communities by ground vegetation composition (smaller participation of lichens) and increase in occurrence of *Betula nana* connected with increase of the air thawing index.

On flat weakly drained top of frost mound the vegetation recovery after the fire is faster than on better drained palsa peatland. Here *Pinus sibirica*- wild rosemary-peat moss-lichen open woodland in 35 years after the fire changed by *Betula nana*-wild rosemary-peat moss-lichen community with *Pinus sibirica* in height 2m.

Stages and rate of vegetation recovery after the fire were revealed.

The ecosystems are established, in which the local temperature decrease observed on a background of the general tendency of temperature increase, caused by dynamics of the vegetation cover.

The carried out researches prove observations of A.P.Tyrtikov (1969), E.B.Belopukhova (1973), V.L.Nevecherya et all. (1975). These researchers marked, that in modern climatic conditions of Western Siberia northern taiga during dynamics of bog vegetation are formed new frost mounds which are considered as some researchers relic formations (Yevseyev, 1976; Brown, Pewe, 1973) for which formation now there are no necessary conditions.

5. Conclusion

In my research the vegetation cover is considered as one of components of the natural ecosystems, closely connected with other components and first of all with soils, underground waters and permafrost for which indication it is used. As the mobile component of ecosystem easily broken at external impact, but capable to self-recovery, vegetation is one of critical components of ecosystems and the major factor of their stabilization.

Long-term monitoring of vegetation cover show that main environmental factors in development of plant communities in the North of West Siberia are water and thermal regime of soil.

Studying of interactions of vegetation with other ecosystem components and revealing of leading factors in vegetation dynamics of region allows more proved to approach to compiling the prediction of vegetation changes in conditions of a varying climate on materials received as a result of long-term monitoring. Use of the interactions existing between the vegetation cover and permafrost, enables to predict on expected tendencies of vegetation development changes of geocryological conditions and to recommend necessary actions on preservation of natural balance in environment.

In all territory of the north of Western Siberia climate changes in time have oscillatory character on a background of the general warming which have begun since 1970th years. On data of Nadym weather station for 1970-2011 the trend to increase of mean-annual air temperature is revealed. Increase of mean-annual temperature has made 0.04^0C in a year.

The steady increase in active layer thickness is connected to rise in air temperature in all natural complexes. Extreme reaction to climatic changes natural complexes of bogs and peatlands in the north of Western Siberia possess. Active layer thickness in palsa peatlands for the 40-years period has increased on 30 %.

Despite of climate warming and observed rise in permafrost temperature single instances of permafrost transition in a thawed condition on all thickness of annual heat turn layer are fixed only.

Author details

Nataliya Moskalenko
Earth Cryosphere Institute SB RAS, Russia

Acknowledgement

The research was supported by Land-Cover Land-Use Change program, project Circumpolar Active Layer Monitoring (CALM, National Science Foundation, Grant NSF OPP-9732051, 0PP-0225603); project Thermal State of Permafrost (TSP, NSF RC-0632400, ARC-0520578) and Council under grant of the President of the Russian Federation (grant NSH-5582.2012.5).

Aleksandar Lazinica is acknowledged for their very useful comments in improving my chapter.

6. References

Andrianov, V.N.; Kozlov, A.N.& Krizuk, L.N. (1973). Engineering-geocryological conditions of pool of middle current   Nadym River. *Trudy of VSEGINGEO*, 62, pp. 79-89

Belopukhova E.B. (1973). Features of modern development of permafrost in Western Siberia. *Proceedings of the Second International Conference on Permafrost*, volume 2, Yakutsk, pp. 84-86

Brown J.E., Pewe T.L. (1973). Distribution of Permafrost in North America and its Relationship to the Environment. *Cl. Review, 1963-1973 "Permafrost" Second International Conference, National Academy of Sciences*. Washington, pp. 15-21

Ershov, E.D; Maksimova, L.N., Medvedev, A.V.& et al. (1994). Reaction of permafrost to global changes of a climate. *Geoecology*, No.5, pp. 11-24

Yevseyev V.P. (1976). Migratory frost mounds in the northeast of the European part and Western Siberia. *Problems of cryolithology*, 5, M.: the Moscow State University, pp. 95-159.

Hollister, R.D.; Webber, P.J. & Tweedie, C.E. (2005). The response of Alaskan arctic tundra to experimental warming: differences between short- and long-term responses. *Global Change Biology*, 11, pp. 1-12

Izrael, J.A.; Pavlov, A.V. & Anokhin, J.A. (2002). Evolution of permafrost zone at modern changes of a global climate. *Meteorology and a hydrology*, No.1, pp.10-18

Kakunov, N.B. & Sulimova, E.I. (2005). Dynamics of geocryological conditions on territory of northern part Timan-Pechora oil-gas province for last 220 years and the forecast of their changes. *Materials of research-and-production conference Problems of engineering - geological maintenance of construction of oil-and-gas complex objects in permafrost zone.* Moscow, FGUP PNIIIS, pp. 107-110

Melnikov, E.S. (ed.) (1983). *Landscapes of Permafrost Zone of the West Siberian Gas Province.* Novosibirsk, Nauka, 166 pp.

Moskalenko, N.G. (1999). *Anthropogenic Vegetation Dynamics in the Permafrost Plains of Russia.* Novosibirsk, Nauka, 280 pp.

Moskalenko, N.G. (ed.) (2006). *Anthropogenic Changes of Ecosystems in West Siberian Gas Province.* Moscow, Earth Cryosphere Institute, 358 pp.

Nelson, F.E.; Lachenbruch, A.H.; Woo, M.-K. & et al. (1993). Permafrost and changing climate. *Permafrost of the Sixth Intern. Conf. Beijing, China,* South China Univ. of Technol. Press., Vol. 2, pp. 987-1005

Nevecherya V.L., Moskalenko N.G. & Tagunova L.N. (1975). About an opportunity of the forecast of change engineering - geocryological conditions depending on character of development of natural complexes at construction in the north of Western Siberia. *Methods of geocryological researches. Trudy of VSEGINGEO,.* 98, M., pp. 16-34.

Oberman, N.G. (2007). Some features of modern degradation of permafrost zone in the Pechora-Ural region. *Materials of International Conference Cryogenic resources of polar regions.* Salekhard, Vol.1, pp. 96-99

Osterkamp, T. & Romanovsky, V. (1999). Evidence for warming and thawing of discontinuous permafrost in Alaska. *Permafrost and Periglacial Processes,* Vol. 10, pp. 17-37

Pavlov, A.V. 1997. The forecast of permafrost zone evolution in the north of Western Siberia (according to monitoring). *Results of basic researches of the Earth permafrost zone in Arctic and Subarctic regions.* Novosibirsk, Nauka, pp. 94-102

Pavlov, A.V. (2008). *Monitoring of permafrost zone.* Novosibirsk, Academy izd-vo "Geo", 229 pp.

Parmuzin, S.J. & Chepurnov, M.B. (2001). Existential dynamics of permafrost in the European North and Western Siberia in XXI century in connection with possible changes of a climate. *Materials of the Second Conference of geocryologists in Russia.* Moscow, the Moscow State University, Vol.2, pp. 231-235

Perlshtein, G.Z.; Pavlov, A.V. & Buiskikh, A.A. (2006). Changes of permafrost zone in conditions of modern climate warming. *Geoecology (engineering geology, hydrogeology, geocryology),* No.4, pp. 305-312

Tyrtikov A.P.(1969). Impact of the vegetation cover on freezing and thawing of grounds. M.: the Moscow State University, , 192 pp.

Tyrtikov, A.P. (1979). *Vegetation Cover Dynamics and Evolution of Permafrost Elements of Relief.* Moscow, Nauka, 115 pp.

Walker, M.D., Wahren, C.H., Hollister, R.D. & et al. (2005). Plant community responses to experimental warming across the tundra biome. *Proceedings of the National Academy of Science of the United States of America (PNAS)* 103(5), pp. 1342-1346

Leibman, M.O.; Moskalenko, N.G.; Orekhov P.T. & et al. (2011). Interrelation of cryogenic and biotic components of geosystems in cryolithozone of the West Siberia along the

"Yamal" transect. V.M.Kotlyakov (editor-in-chief). *Polar Cryosphere and continental waters.* Paulsen Editions. Moscow-Saint-Petersburg, pp.171-192

Climate Change Assessment Due to Long Term Soil Moisture Change and Its Applicability Using Satellite Observations

Hui Lu, Toshio Koike, Tetsu Ohta,
Katsunori Tamagawa, Hideyuki Fujii and David Kuria

Additional information is available at the end of the chapter

1. Introduction

In researches related to the climate change, soil moisture is serving as an excellent environmental indicator controlling and regulating the interaction between the atmosphere and the land surface (Betts et al. 1996; Entekhabi et al. 1996). Soil moisture controls the ratio of runoff and infiltration (Delworth & Manabe 1988; Wagner et al. 2003), decides the energy fluxes (Entekhabi et al. 1996; Prigent et al. 2005) and influents vegetation development and then carbon cycle (Pastor & Post 1986; Melillo et al. 2002). Moreover, soil moisture is also an important factor in animal and plant productivity and it can even constrain to the interaction between natural system and anthropic activity. Therefore, the distribution pattern of soil moisture, both spatial and temporal, is the key of climate system modelling. Moreover, a long term soil moisture data set on a region scale therefore could provide valuable information for researches such as climate change and global warming (Seneviratne et al. 2006), and then improve the weather forecasting (Beljaars et al. 1996; Schar et al. 1999) and water resources management.

Traditionally, soil moisture is measured at point scale through filed samplings and/or automatic instruments, for example, Time Domain Reflector (TDR). These methods are commonly used to provide accurate and continuous soil moisture information and adopted by the meteorology, hydrology and agriculture stations. Through such point scale measurement, long term and accurate soil moisture information was collected. Such information is useful in climate change study, but it is only at point scale and is only available at limited locations.

Contrary to the limitations of *in situ* measurements, numerical models can provide continuous estimates of soil moisture over the entire soil profile at any scale. Therefore, the numerical simulation, in which meteorological observations and land surface models are employed, became the main approach for studying soil moisture variation. However, the limitations of the numerical simulation approach such as the simplification of physical processes, the uncertainty in parameters and the biases in atmospheric data are significant. Consequently, the accuracy of simulation results is highly dependent on the quality of the atmospheric data from which the uncertainties and biases were inherited. ·

Unlike model simulations, remote sensing is able to provide land surface soil moisture observations that do not rely on atmospheric variables. This independence is very important to ensure objectivity of the climate change impact assessment. Satellite remote sensing offers a possibility to measure surface soil moisture at regional, continental and even global scales. Although surface soil moisture can be estimated indirectly from visible/IR remote sensing data (Verstraeten et al. 2006; Gillies & Carlson 1995), it failed to produce routinely soil moisture map mainly due to factors inherent in optical remote sensing, such as atmosphere effects, cloud masking effects and vegetation cover masking effects. Currently, the most popular technique is the microwave remote sensing which provides a means of direct surface soil moisture measurement for a range of vegetation cover conditions (Oh, Sarabandi & Ulaby 1992; Kerr et al. 2001; Paloscia et al. 2008; Njoku & Chan 2006). The scientific basis of microwave soil moisture remote sensing is the strong relationship between the soil dielectric properties and its liquid moisture content (Hipp 1974; Wang & Schmugge 1980). Moreover, extra advantages of microwave remote sensing include: (1) long wavelength in microwave region which enable the low frequency microwave signals to penetrate clouds and to provide physical information of the land surface; and (2) independent of illumination source which enables the spaceborne sensors to observe earth all-day with all-weather coverage.

There are two approaches through which microwave remote sensing estimates surface soil moisture: active ways by Radar and/or SAR with high spatial resolution (in the order of ten to hundred meters) and long revisiting period (about 1 month), passive ways by radiometers with coarse resolution (~ order of 10 km) and frequent temporal coverage (daily or bi-daily). Considering the temporal resolution requirement of the meteorological and hydrological modelling, passive ways are more suitable for the application in these fields. Passive microwave remote sensing has been recognized as a potential method for measuring soil moisture with a large spatial coverage, while the sensors operated at low frequencies have been acknowledged to be capable to estimate soil moisture reliably (Njoku & Li 1999; Wigneron et al. 1998; Njoku & Entekhabi 1996). For instance, the Advanced Microwave Scanning Radiometer for the Earth Observing System (AMSR-E) is believed to offer state-of-the-art soil moisture estimation through the combination of low frequency observations at 6.9, 10.65 and 18.7 GHz (Paloscia, Macelloni & Santi 2006; Njoku et al. 2003). In terms of soil moisture temporal distribution, the Special Sensor Microwave/Imager (SSM/I) equipped on Defense Meteorological Satellite Program (DMSP) satellites, measuring the brightness temperature of the earth at 19.35, 22.235, 37 and 85.5 GHz with a history around 20 years, is highly expected to provide long-term global soil moisture estimation (Paloscia et al. 2001; Prigent et al. 2000; Jackson 1997; Hollinger et al. 1987). Before SSM/I, the Scanning Multichannel Microwave radiometer (SMMR) on board Nimbus-7 Pathfinder (Gloersen & Barath 1977), provides brightness temperature observation at 6.6, 10.7, 18.0, 21 and 37 GHz from 1978-1987.

In this study, we investigated the applicability of the remotely sensed soil moisture in climate change studies. We retrieved long term soil moisture over the whole Africa continent from SSM/I by using a revised algorithm. The long term soil moisture data set was analyzed to demonstrate the advantages of passive microwave remote sensing in supporting the climate change studies. The results were compared with the reanalysis data subset from ECMWF (European Centre for Medium-Range Weather Forecasts) 40-year Reanalysis data (ERA-40) and The National Centers for Environmental Prediction/National Center for Atmospheric Research (NCEP/NCAR) reanalysis data.

The paper is organized as follows. In Section 2 we describe the structure of the soil moisture retrieving algorithm. Section 3 presents a short introduction to the study region and used data.Section 4 discusses the application case by using SSM/I soil moisture data over Africa. Section 5 ends this paper with some concluding remarks.

2. Soil moisture retrieval algorithm

The soil moisture retrieval algorithm for passive microwave remote sensing data consists of three parts: a forward model, a retrieving algorithm and the ancillary data set.

2.1. The forward model: radiative transfer model

Our algorithm is based on a look up table, which is a database of brightness temperature simulated by a radiative transfer model for various possible conditions. The quality of retrieved soil moisture, therefore, is heavily dependent on the performance of the radiative transfer model. So, the main task of our algorithm development is to develop a physically-based soil moisture retrieval algorithm, which is able to estimate soil moisture content from low frequency passive microwave remote sensing data and to overcome the misrepresent problems occurred in dry areas.

2.1.1. The radiative transfer process for land surface remote sensing

For the land surface remote sensing by spaceborne microwave radiometers, the radiative transfer process from land to space can be divided into as four stages as follows:

1.   Radiative transfer inside soil media

The initial incident energy is treated as the one starting from the deep soil layer, which propagates through many soil layers, attenuating by the soil absorption effects (dominative at wet cases) and volume scattering effects (dominative at dry cases), experiencing multi-reflection effects between the interfaces of soil layers, finally reaching the soil/air interface.

2. Surface scattering process at soil/air interface

At the soil/air interface, the surface scattering influences this upward initial radiation by changing its direction, magnitude and polarization status. At the same time, the downward radiation from the cosmic background, atmosphere, precipitation and canopy are reflected by the air/soil interface, and parts of the reflected radiation propagate along the same direction as that emitted from the soil layers.

The upward radiation just above the soil/air interface, therefore, is not only the product of soil medium but also the product of downward radiation.

3. Radiative transfer inside vegetation layers

After leaving the soil/air interface, the upward radiation propagates through the canopy layer (if there are vegetations), experiences the volume scattering effects from the leaves and stems of vegetations and the multi-reflection effects between canopy/air and soil/air interfaces. At the same time, parts of the upward radiation from vegetations join our target radiation.

4. Radiative transfer inside atmosphere layers

After transmitting from vegetation layer, the radiation continues its way, traversing the cloud and precipitation layers, affected by the absorptive atmosphere gases, scattered by precipitation drops, incorporating the emission from surroundings, finally detected by the sensors boarded on satellites.

The story of radiative transfer is so complicate that make it necessary to simplify the process to make it computable. In microwave region, the reflectivity of the air/soil interface is generally small. The downward radiation from vegetation and rainfall, which is reflected by the soil surface, therefore, is neglected. Moreover, for the lower frequencies region of microwave, the atmosphere is transparent. Finally, after neglecting all the downward radiation and parts of upward radiation from surroundings, the radiative transfer model is written as:

$$T_b = T_{bs}e^{-\tau_c}e^{-\tau_r} + (1-\omega_c)(1-e^{-\tau_c})T_ce^{-\tau_r} + \int(1-\omega_r(R))(1-e^{-\tau_r(R)})T_r(R)dR \tag{1}$$

where T_{bs} is the emission of the soil layer, T_c is the vegetation temperature, T_r is the temperature of precipitation droplets, τ_c and ω_c are the vegetation opacity and single scattering albedo, and τ_r and ω_r are the opacity and single scattering albedo of precipitation.

For the frequencies less than 18GHz, equation (1) can be even simplified by omitting the precipitation layer, as:

$$T_b = T_{bs}e^{-\tau_c} + (1-\omega_c)(1-e^{-\tau_c})T_c \tag{2}$$

2.1.2. Radiative process inside soil media: profile effects and volume scatting effects

Microwave can penetrate into soil media, especially for dry cases, in which the penetration depth of C-band is about several centimeters. The soil moisture observed by microwave remote sensing, therefore, is inside a soil media with a volume of several centimeters depth. The radiative transfer process inside a soil media includes various effects, such as moisture and temperature profile effects and the volume scattering effects of dry soil particles. To simulate these effects, the dielectric constant model should be addressed at first.

1.   Dielectric constant model of soil

In the view of microwave, soil is a multi-phase mixture, with a dielectric constant decided by moisture content, bulk density, soil textural composition, soil temperature and salinity. In our algorithm, the dielectric constant of soil is calculated using Dobson model (Dobson et al. 1985):

$$\varepsilon_{soil}^{\alpha} = 1 + \frac{\rho_b}{\rho_{ss}}(\varepsilon_{ss}^{\alpha} - 1) + m_v^{\beta}\varepsilon_{fw}^{\alpha} - m_v \qquad (3)$$

where ρ_b is the bulk density of sand, ρ_{ss} = 2.71 is the density of solid sand particle; ε_{ss} = (4.7, 0.0) is the dielectric constant of sand particle; m_v is the volumetric water content; ε_{fw} is the dielectric constant of free water, can be calculated by the model proposed by Ray (1972); $\alpha=0.65$ is an empirical parameter; and β is a soil texture dependent parameter as follows:

$$\beta = 1.09 \quad 0.11S \mid 0.18C \qquad (4)$$

where S and C are the sand and clay fraction of the soil, respectively.

2.   Profile effects of soil media

The heterogeneity inside soil media causes the so-called profile effects. The profile effects can be accounted for by using the simple zero-order noncoherent model proposed by Schmugge and Choudhury (1981) or by more complicate first-order noncoherent model given by Burke et al. (1979). The volume scattering effects inside soil media are not included in both models.

In order to include the volume scattering effects, a more complicate model was adopted in our algorithm. We assumed that the soil has a multi-layer structure and is composed of many plane-parallel and azimuthally symmetric soil slabs with spherical scattering particles. The radiative transfer process in a plane-parallel and azimuthally symmetric soil slab with spherical scattering particles can be expressed as (Tsang & Kong 1977):

$$\mu\frac{d}{d\tau}\begin{bmatrix} I_v(\tau,\mu) \\ I_h(\tau,\mu) \end{bmatrix} = \begin{bmatrix} I_v(\tau,\mu) \\ I_h(\tau,\mu) \end{bmatrix} - (1-\omega_0)B(\tau)\begin{bmatrix} 1 \\ 1 \end{bmatrix} - \frac{\omega_0}{2}\int_{-1}^{1}\begin{bmatrix} P_{VV} & P_{VH} \\ P_{HV} & P_{HH} \end{bmatrix}\begin{bmatrix} I_v(\tau,\mu') \\ I_h(\tau,\mu') \end{bmatrix}d\mu' \qquad (5)$$

where $I_P(\tau,\mu)$ is the radiance at optical depth τ ($d\tau=K_e dz$, with extinction coefficient K_e and layer depth dz) in direction μ for polarization status P (horizontal or vertical), ω_0 is the single scattering albedo of a soil particle, $B(\tau)$ is the Planck function and P_{ij} ($i, j=H$ or V) is the scattering phase function. The 4-stream fast model proposed by Liu (1998) solves (5) by using the discrete ordinate method and assuming that no cross-polarization exist. The

Henyey-Greenstein formula (Henyey & Greenstein 1941) is used to express the scattering phase function.

3.   Volume scattering effects of dry soil particles

With considering the facts that the soil particles are densely compacted, the multi-scattering effects of soil particles should be accounted for. In our algorithm, this volume scattering effect was calculated by the so-called dense media radiative transfer theory (DMRT) under Quasi Crystalline Approximation with Coherent Potential (QCA-CP) (Wen et al. 1990; Tsang & Kong 2001). Dense Media radiative transfer theory was derived from Dyson's equation under the quasi-crystalline approximation with coherent potential (QCA-CP) and the Bethe-Salpeter equation under the ladder approximation of correlated scatterers.

By using the DMRT, the extinction coefficient K_e and albedo ω used in equation (5) were calculated. And then the radiance of each soil slab was calculated by the 4-stream fast model. The radiance just below the soil/air interface was obtained by integrating the radiance from bottom layer to the top layer. Finally, the apparent emission of soil media, T_{bs} in equation (1) and (2), was obtained.

2.1.3. Surface roughness effects

When an electromagnetic wave reaches the air/soil interface, it suffers the reflection and refraction due to the dielectric constant changing in the two sides of the interface. The roughness of the interface divides the reflected wave into two parts, one is reflected in the specular direction and another is scattered in all directions. Generally, the specular component is often referred to as the coherent scattering component. And the scattered component is known as the diffuse or noncoherent component, which consists of power scattered in all directions but with a smaller magnitude than that of the coherent component. Qualitatively, Surface roughness increases the apparent emissivity of natural surfaces, which is caused by increased scattering due to the increase in surface area of the emitting surfaces. And it was demonstrated by many researches that the surface roughness has a nonnegligible effects on the accuracy of soil moisture retrieval by spaceborne microwave sensors (Oh & Kay 1998; Singh et al. 2003). In general, the surface roughness effects are simulated by two ways: semi-empirical models and fully physical-based models.

1.   Semi-empirical models

The semi-empirical models are simply and do not cost too much computation efforts. The parameters used in semi-empirical models are often derived from field observations. Depending on the parameters involved, three different semi-empirical models: Q-H model (Choudhury et al. 1979; Wang & Choudhury 1981), Hp model (Mo & Schmugge 1987; Wegmuller & Matzler 1999; Wigneron et al. 2001) and Qp model (Shi et al. 2005).

2.   Fully physical-based model

In our algorithm, we simulated the land surface roughness effect using the Advanced Integral Equation Model (AIEM) (Chen et al. 2003). AIEM is a physically-based model with only two parameters: standard deviation of the height variations σ (or *rms* height) and surface correlation length l. AIEM is an extension of the integral Equation Model (IEM) (Fung, Li & Chen 1992). It has been demonstrated that IEM has a much wider application range for surface roughness conditions than other models such as the Small Perturbation Model (SPM), Physical Optics Model (POM) and Geometric Optics Model (GOM). AIEM improves the calculation accuracy of the scattering coefficient compared with IEM by retaining the absolute phase term in the Green's function.

By coupling AIEM with DMRT (DMRT-AIEM), this radiative transfer model for soil media is fully physically-based. As such, the parameters of DMRT-AIEM, such as the *rms* height, correlation length and soil particle size, have clear physical meanings and their values can be obtained either from field measurement or theoretical calculation.

2.1.4. Vegetation masking effects

The existence of canopy layers complicates the electromagnetic radiation which is originally emitted solely by soil layers. The vegetation may absorb or scatter the radiation, but it will also emit its own radiation. The effects of a vegetation layer depend on the vegetation opacity τ_c and the single scattering albedo of vegetation ω_c (Schmugge & Jackson 1992). The vegetation opacity in turn is strongly affected by the vegetation columnar water content W_c. The relationship can be expressed as (Jackson & Schmugge 1991):

$$\tau_c = \frac{b'\lambda^\chi W_c}{\cos\theta} \tag{6}$$

where λ is the wavelength, θ the incident angle, W_c the vegetation water content. χ is 1.38 and b' is 9.32.

The single scattering albedo, ω_c, describes the scattering of the emitted radiation by the vegetation. It is a function of plant geometry, and consequently varies according to plant species and associations. The value of it is small in the low frequency microwave region (Paloscia & Pampaloni 1988; Jackson & Oneill 1990). In our algorithm, ω_c is calculated by

$$\omega_c = \omega_0 \cdot \sqrt{W_c} \tag{7}$$

The value of albedo parameter ω_0 is decided empirically in current researches. Experimental data for this parameter are limited, and values for selected crops have been found to vary from 0.04 to about 0.12.

By combing the T_{bs} solved by equation (5), the surface reflectivity calculated by AIEM, the vegetation opacity τ_c calculated by equation (6), and the vegetation single scattering albedo ω_c estimated by equation (7), a physical-based radiative transfer model was developed.

2.2. The algorithm

The basis of our algorithm is a database of brightness temperature and/or some indexes calculated from brightness temperature. By searching the data base (or look up table) with the satellite observation as the input, soil moisture and other related variables of interest can be estimated quickly. Such high searching speed is the main reason why we adopt the look up table method for soil moisture retrieval. The implementation of our algorithm consists of three steps: (1) Fixing the parameters used in the forward model; (2) Generating a look up table by running forward model; and (3) retrieving soil moisture by searching the look up table.

2.2.1. Parameterization

As in other physically-based algorithms, such as that developed by Njoku et al. (2003) and the single channel algorithm developed by Jackson (1993), the parameters used in our algorithm have clear physical meanings. This advantage derives from the strength of the forward radiative transfer model. Before running the forward RTM to generate look up table, the parameters should be confirmed at first. The parameters to be confirmed include *rms* height (h), correlation length (l), soil particle sizes (r) and vegetation parameters such as χ and b'. Currently, we can obtain these parameters through the best-fitting method.

For the region where in-situ soil moisture and temperature observation are available and when such observation are also representative, we can use a best-fitting way to optimize parameters. In order to simplify the calculation, low frequencies simulation and observation were used. These parameters are optimized by minimizing the cost function:

$$J(h,l,r,b',\chi,...) = \sum_{i=1}^{n}\sum_{f}\sum_{p=H,V} ABS[TB_{sim}(i,f,p) - TB_{obs}(i,f,p)] \tag{8}$$

where the subscript *sim* denotes the model simulated value and *obs* is the observed value. n is the number of samples used in the optimization. p denotes the polarization status: H for horizontal polarization and V for vertical. f is some frequency in the long wavelength region where the atmospheric effect may be ignored, such as 6.9, 10.7 and 18.7 GHz of AMSR-E, 1.4 GHz of SMOS and 19GHz of SSM/I.

2.2.2. Look up table generation

After Step 1, the optimal parameter values are then stored in the forward RTM. We then run the forward model by inputting all possible values of variables used in Equation (1), such as soil moisture content, soil temperature, vegetation water content and atmosphere optical thickness. A family of brightness temperatures is then generated. Based on this brightness temperature database, we select brightness temperatures of special frequencies and polarization to compile a lookup table or to calculate some indices to compile a lookup table. For example, in order to partly remove the influences of physical temperature, the ratio of brightness temperature at different frequencies and polarizations can be used. For instant, we can compile a look up table by using the index of soil wetness (ISW) (Koike et al. 1996; Lu et al. 2009), and Polarization Index (PI) (Paloscia & Pampaloni 1988).

$$PI(19) = \frac{TB(19,V)-TB(19,H)}{TB(19,V) + TB(19,H)} \tag{9}$$

$$ISW(19,37) = \frac{TB(37,H)-TB(19,H)}{TB(37,H)+TB(19,H)} \tag{10}$$

where TB(19, V) and TB(19, H)and represents the brightness temperature observed at 19GHz vertical and horizontal polarization channel, respectively, TB(37,H) is the observation at 37GHz horizontal polarization channel.

2.2.3. Soil moisture estimation

The lookup table generated in Step 2 is reversed to give a relationship which maps the brightness temperature or indices obtained from satellite remote sensing data to the variables of interest (such as soil moisture, soil temperature and vegetation water content). Finally, we estimate soil moisture content by linear interpolation of the brightness temperature or indices into the inverted lookup table.

2.3. Ancillary data sets

In order to run forward model, various parameters are needed. These parameters are provided from ancillary data set. Soil parameters, including soil texture and bulk density are from FAO data set. Vegetation parameters used in calculated optical thickness are from references and vegetation types. Land cover map is also used to identify the regions with dense vegetation and big water surface, where the current soil moisture retrieval algorithm is not applicable.

3. Study regions and used data

3.1. Study regions

Africa continent, comparing to the other continent in the world, is less impacted by industry development. There is the largest desert of the world and the arid-like climate is dominative in the most part of Africa continent. Therefore, Africa continent has an environment vulnerable to climate change. On the other hand, it is very difficult to monitor the environmental changes in this region due to its tough situations and less-development.

In this study, the application region is covered the whole Africa continent, from 20W to 60E and from 40S to 40N. The application period is from 1988 to 2007, which is the period covered by SSM/I observation.

3.2. Reanalysis data

In this study, due to the limitation of in situ observation data in Africa, only the numerical model output is available for the results inter-comparison. The current general circulation models (GCM) are mainly focusing on the atmospheric variable simulation. However, the surface soil moisture simulations are not as reliable as atmospheric variables (Li, Robock &

Wild 2007). From a view of water budget, the land surface water comes from atmosphere, controlled by the precipitation and radiation. Consequently, we use the atmospheric variables from the reanalysis data to express the land surface wetness change. The variable used in representing climate change tendency is the net water flux at the land surface, which represents the vertical water budget in the atmosphere as:

$$W_{net} = E - P \tag{11}$$

where W_{net} is the net water flux, E is the evaportranspiration flux and P is the precipitation flux. The positive value of net water flux means drying situation, while the negative value represents wetting trends. In order to partly mitigate the bias inherited in the individual GCM, two sets of reanalysis data, ERA-40 and NCEP, were used in this study.

ERA-40 is a reanalysis of meteorological observations produced by the European Centre for Medium-Range Weather Forecasts (ECMWF) (Uppala et al. 2005). The ERA-40 has a TL159 horizontal resolution and 60 vertical levels. The ERA-40 dataset used in this study has been obtained from the ECMWF data server (http://data.ecmwf.int/data/d/era40_daily/).

The National Centers for Environmental Prediction/National Center for Atmospheric Research (NCEP/NCAR) reanalysis (which is referred to as the NCEP reanalysis hereafter) is a global reanalysis that was initiated in 1948 and continues to the near real-time at a horizontal resolution of T62 and a temporal resolution of 6 hours (Kalnay et al. 1996). The dynamic model that is used in the NCEP reanalysis is the NCEP global spectral model (Kalnay, Kanamitsu & Baker 1990; Karamitsu 1989; Kanamitsu et al. 1991), in which the land surface processes are treated as portions of the atmospheric model in a relatively simple approach.

3.3. SSM/I remote sensing data

The brightness temperature of SSM/I observed at 19.35 GHz and 37GHz was used as the observation data in this research. The SSM/I is flown by the Defense Meteorological Satellite Program (DMSP) on two operational polar orbiting platforms. The first spacecraft of SSM/I series (F08) was launch in June 1987. Currently working one is the F17 launch in December 2006. TABLE I gives the launch date and end date of SSM/I satellites.

The nadir angle for the Earth-viewing reflector of SSM/I is 45°, which results in an Earth incidence angle of 53.4° ± 0.25°. The lower frequency channels (19, 22, and 37 GHz) are sampled so that the pixel spacing is 25 km, and the 85 GHz channels are sampled at 12.5 km pixel spacing. More information on the SSM/I can be found in (Hollinger et al. 1987). The brightness temperature data of SSM/I is archived in the National Snow and Ice Data Center (NSIDC) (Armstrong et al. 2012). The brightness temperature data used in this research can be found at the following webpage: http://nsidc.org/data/docs/daac/nsidc0032_ssmi _ease_tbs.gd.html.

Satellite	Launch Date	End Date
F08	July 1987	December 1991
F10	December 1990	November 1997
F11	December 1991	May 2000
F13	May 1995	November 2009
F14	May 1997	August 2008
F15	December 1999	
F16	October 2003	
F17	December 2006	

Table 1. SSM/I Launch Dates and End Dates

4. Results and discussions

Using the algorithm, daily soil moisture was retrieved from SSM/I TB data. The daily soil
moisture is then converted into monthly average soil moisture data set. Figure 1 shows the
monthly averaged soil moisture data at July of 1988.

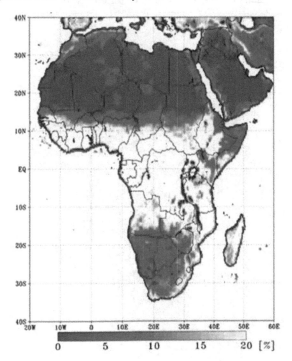

Figure 1. Retrieved montly averaged soil moisture from SSM/I

From Figure 1, it is clear that the retrieved soil moisture over Africa is less than 20%., which may be an underestimation. But the distribution patterns are in a realistic ways: (1) coastal regions generally show larger values than inland regions; (2) the desert regions have the smallest values; (3) the South Africa is wetter than North Africa, because July is the winter month and raining season of South Africa.

In central Africa regions, retrieval algorithm failed to estimate soil moisture values from SSM/I observation. It is because the vegetation is very dense in these regions and the microwave signal could not penetrate the canopy layers and failed to reach the soil layer. For such regions with dense vegetation, longer wavelength channel observation is needed to detect land surface information.

4.1. Climate change tendency

In order to identify the climate change tendency, long term averaged data were used. We first divided 20 years soil moisture data into two groups: the first decade from 1988 to 1997 and the second decade from 1998 to 2007.And then we calculated the decade-average summer soil moisture for each decade. It means ten years average of JJA (June, July, and August) for northern hemisphere and of DJF (December, January, and February) for southern hemisphere. Finally, we calculated the difference between the second decade-average and the first decade-average.

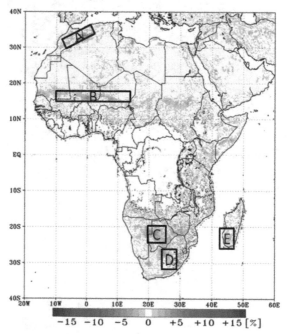

Figure 2. Difference between two decade-average summer soil moisture (%)

Figure 2 shows the difference between (1998~2007) and (1988~1997). From this figure, we can identify some climate change tendencies over the Africa continent:

1.   There is a drying tendency in the coastal regions of Mediterranean Sea:

These regions include: Morocco, Algeria, Tunisia, and Libya. The drying tendency in region A is most remarkable. This finding is in agreement with the report of Esper et al. (2007): the drought occurrence frequency in Morocco is increasing during last decade.

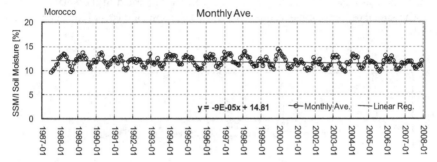

Figure 3. Monthly averaged soil moisture at region A from 1987-2008

Figure 3 shows the monthly averaged soil moisture time series for the region A in figure 2. It represents the area-averaged soil moisture variation in Morocco. A linear regression line is also shown in figure 3. The inter-annual variation of soil moisture is not obvious while the maximum values of each year decreases obviously. The regression slope for the region A (-0.00009) is less than zero. It means the surface soil moisture in this region decreases constantly.

2.   There is a wetting tendency in the belt region from 14N to18N

It is the south boundary of the Sahara desert, and its land cover type is grassland. Countries located in this region are: Mauritania, Mali, Niger, Chad, Senegal and Sudan. The wetting tendency in region B as marked in the figure is the most remarkable in this belt.

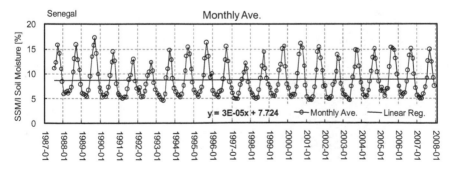

Figure 4. Monthly averaged soil moisture at region B from 1987-2008

Figure 4 shows the monthly averaged soil moisture time series in Senegal. It is clear form figure 4 that the second decade is wetter than the first decade, with less inter-annual varation. The linear regression slope in figure 4 (0.00003) is larger than 0, meaning that the surface soil moisture increases constantly during 1987-2008.

3.   There is a general wetting tendency in the South Africa region

Countries showing such tendency include: Namibia, Botswana, Zimbabwe, and South Africa, marked as regions C and D in figure 2. Region C is located in the Kalahari Desert, which showing most remarkable wetting tendency. The main land cover type of region C is savanna.

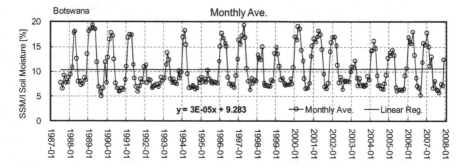

Figure 5. Monthly averaged soil moisture at region C from 1987-2008

Figure 5 shows the soil moisture time series in Botswana of region C. During the first decade (1988-1997), there are three years with extreme dry situation, while there only one year during the second decade. The linear regression slope of region C is larger than zero and it means soil moisture increases in this region for the last 20 years. It is mainly a result of less extreme dry years in the second decade.

4.   There is a slight wetting tendency in the southern part of Madagascar Island

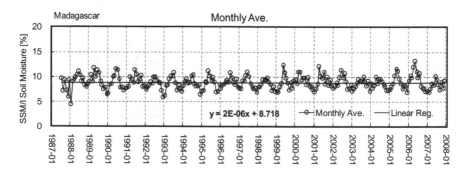

Figure 6. Monthly averaged soil moisture at region E from 1987-2008

As shown in figure 6, there are four wet anomaly years in the second decade, i.e. 1999, 2001, 2005 and 2006. The linear regions slope of this region is also larger than zero. It means the soil moisture in region E increases during last 20 years and it mainly due to more anomaly wet years in recent 10 years.

4.2. Comparison to ERA-40

Figure 7 presents 20 years averaged summer (*E-P*) over the continent, calculated from ERA-40 reanalysis data. Comparing with Figure 2, the same wetting tendency in region B, C, D and E can be identified easily. The findings from remote sensing data are in good agreement with the reanalysis data.

As shown in Figure 7, the wetting tendency is remarkable in the tropical region. But current remote sensing data failed to provide soil surface information in this region. We cannot identify any remarkable drying tendency in the coast region along Mediterranean Sea.

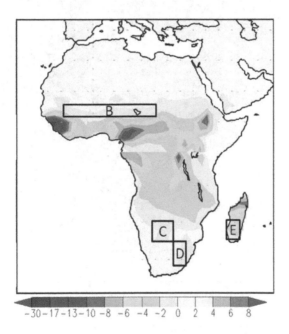

Figure 7. 20 years averaged (E-P) from ERA-40 (unit: mm)

It means that the drying trend in Morocco is not significant in ERA-40 data. By using only ERA-40, the land surface wetness change trends in Morocco will be missed. It means that remote sensing data is able to provide complementary climate change information to the traditional reanalysis data. Moreover, the climate change trend derived from ERA-40 shows a wet bias over the whole Africa continent.

4.3. Comparison to NCEP

As discussed in the introduction part, all numerical models have limitation. In order to overcome the shortage of individual model, inter-comparison of multi-model results were proposed in this research. Beside ERA-40, NCEP reanalysis data was also used to check the performance of remote sensing results.

Figure 8 shows the 20 years averaged net water flux over Africa by using NCEP reanalysis data. Comparing Figure 8 with Figure 2, both remote sensing data and NCEP data show the drying trends in region A and the wetting trends in region D and E. But NCEP data failed to represent the wetting trends in region B and C. Comparing Figure 8 with 7, both the area and absolute values of the wetting regions of NCEP data is much smaller than those of ERA-40. It suggests that the climate change trend derived from NCEP data shows a dry bias over the continent.

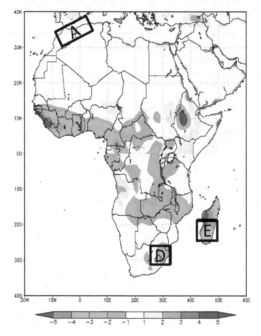

Figure 8. Long-term net water flux over Africa from NCEP (unit:mm)

5. Conclusions

Spatial distributed soil moisture information is an essential parameter for hydrological, meteorological and ecological studies. In this paper, we developed a 20-year soil moisture data set over Africa continent by using a soil moisture retrieval algorithm which estimated soil moisture from passive microwave remote sensing data provided by SSM/I.

Using this long-term soil moisture data set, some climate change tendencies were indentified:

- There is a remarkable drying trend in the north part of Morocco and Libya, which is verified by local observation. The surface soil moisture in this region decreases constantly with small inter-annual variations during last 20 years. This drying trend is also captured by the NCEP reanalysis data.
- There is a wetting belt in the south boundary of Sahara desert, including Mauritania, Mali, Niger, Chad, Senegal and Sudan. The surface soil moisture in this region increases constantly with small inter-annual variation during past 10 years. This wetting trend is also captured by the ERA-40 reanalysis data.
- There is a general wetting tendency in South Africa region, including the Kalahari Desert. From the viewpoint of land cover, grass land and savanna region is getting wet. Such wetting tendencies are in good agreement with the analysis results of ERA-40 reanalysis data. NCEP reanalysis data also shows a wetting trend in parts of this domain (region D). The wetting trend in this region is mainly a result of less extreme dry years in the second decade.
- There is a wetting trend in the south of Madagascar Island, which is also captured by both ERA-40 and NCEP data. The wetting trends in the island it mainly due to more anomaly wet years in recent 10 years.

This study demonstrates that passive microwave remote sensing data is able to provide independent and complementary land surface information to the climate change research. It is a kind of "observation" data, which does not rely on any model assumption and initial conditions. As shown in the results from ERA-40 and NCEP, model simulations have some biases and fail to capture all climate change information over the whole continent. Remote sensing data, therefore, could provide independent and complementary information in climate change study.

But there are some limitations in current remote sensing data, for example, it cannot provide soil information in dense vegetation region. This shortage can be partly overcome as the launch of Soil Moisture and Ocean Salinity mission (SMOS) (Kerr et al. 2001) and future coming Soil Moisture Active and Passive mission (SMAP) (Entekhabi et al. 2008). L band microwave observation which has a longer wavelength will be available and the shadowing effects of vegetation could be gradually alleviated. In addition to the adoption of new sensors, merging remote sensing data and land surface models into a Land Data Assimilation System (LDAS) (Reichle & Koster 2005) is highly expected to the maximum usage of remote sensing data and our knowledge of climate systems.

The data used in this study covers the period from 1987 to 2008. The soil moisture spatiotemporal variation characteristics derived from this research are therefore just from a short period of 20 years. By integrating the remote sensing observations made by SMMR, SSM/I, Tropical Rainfall Measuring Mission Microwave Imager (Jackson & Hsu 2001), AMSR-E and oncoming Global Change Observation Mission – Water (Imaoka et al. 2010) with the proposed ensemble method, a long term soil moisture time series beginning in 1978 can be reconstructed. Such a long-term remote sensing data set has high potential in the assessment of global change impacts on water resources, agriculture, and ecology.

Author details

Hui Lu
Ministry of Education Key Laboratory for Earth System Modeling, and Center for Earth System Science, Tsinghua University, Beijing, China

Toshio Koike, Tetsu Ohta and Katsunori Tamagawa
The Department of Civil Engineering, The University of Tokyo, Tokyo, Japan

Hideyuki Fujii
Earth Observation Research Center, Japan Aerospace Exploration Agency, Ibaraki, Japan

David Kuria
Geomatic Engineering and Geospatial Information Science Department, Kimathi University College of Technology, Kenya

Acknowledgement

This work was jointly supported by the National Natural Science Foundation of China (No. 51109111 and No. 51190092) and Tsinghua University Initiative Research Program (No. 2011081132).

6. References

Armstrong RL, Knowles KW, Brodzik MJ, Hardman MA 2012, DMSP SSM/I-SSMIS Pathfinder Daily EASE-Grid Brightness Temperatures. Boulder, Colorado USA: National Snow and Ice Data Center. Digital media.

Betts AK, Ball JH, Beljaars ACM & Viterbo PA 1996, 'The land surface-atmosphere interaction: a review based on observational and global modeling perspectives', *Journal of Geophysical Research*, vol. 101, pp. 7209–7225.

Beljaars ACM, Viterbo P, Miller MJ & Betts AJ 1996, 'The anomalous rainfall over the United States during July 1993: sensitivity to land surface parameterization and soil moisture anomalies', *Monthly Weather Review*, vol. 124, pp. 362-383.

Burke, WJ, Schmugge TJ & Paris JF 1979, 'Comparison of 2.8 and 21 cm microwave radiometer observations over soils with emission model calculation', *Journal of Geophysics Research*, vol. 84, pp. 287-294.

Chen KS, Wu TD, Tsang L, Li Q, Shi JC & Fung AK 2003, 'Emission of rough surfaces calculated by the integral equation method with comparison to three-dimensional moment method simulations', *IEEE Transactions on Geoscience and Remote Sensing*, vol. 41, pp. 90-101.

Choudhury BJ, Schmugge TJ, Chang ATC & Newton NR 1979, 'Effect of surface roughness on the microwave emission from soils', *Journal of Geophysical Research*, vol. 84, pp. 5699-5706.

Delworth T & Manabe S 1988, 'The influence of potential evaporation on the variability of simulated soil wetness and climate', *Journal of Climate*, vol. 13, pp. 2900–2922.

Dobson MC, Ulaby FT, Hallikainen MT & Elrayes MA 1985, 'Microwave dielectric behavior of wet soil 2: dielectric mixing models', *IEEE Transactions on Geoscience and Remote Sensing*, vol. 23, pp. 35-46.

Entekhabi D, Njoku E, O'Neill P, Spencer M, Jackson T, Entin J & Kellogg Elm K 2008, 'The Soil Moisture Active/Passive Mission (SMAP)', *IEEE Geoscience and Remote Sensing Symposium 2008 (IGARSS08)*, vol. 3, pp. 1-4.

Entekhabi D, Rodrigues-Iturbe I & Castelli F 1996, 'Mutual interaction of soil moisture state and atmospheric processes', *Journal of Hydrology*, vol. 184, pp. 3-17.

Esper J, Frank D, Bu"ntgen U, Verstege A, Luterbacher J & Xoplaki E 2007, 'Long-term drought severity variations in Morocco', *Geophysical Research Letters*, vol. 34, pp.1-5.

Fung AK, Li ZQ & Chen KS 1992, 'Backscattering from a randomly rough dielectric surface.', *IEEE Transactions on Geoscience and Remote Sensing*, vol. 30, pp. 356-369.

Gillies RR & Carlson TN 1995, 'Thermal remote sensing of surface soil water content with partial vegetation cover for incorporation into climate models', *Journal of Applied Meteorology*, vol. 34, pp. 745-756.

Gloersen, P & Barath FT 1977, 'A Scanning Multichannel Microwave Radiometer for Nimbus-G and SeaSat-A', *IEEE Journal of Oceanic Engineering*, vol. 2, pp.172-178.

Henyey LC & Greenstein JL 1941, 'Diffuse radiation in the galaxy', *The Astrophysics Journal*, vol. 93, pp. 70–83.

Hipp J.E. 1974, 'Soil electromagnetic parameters as functions of frequency, soil density, and soil moisture', *Proceedings of the IEEE*, vol. 62, pp.98–103

Hollinger J, Lo R, Poe G, Savage R & Pierce J 1987, 'Special Sensor Microwave/Imager user's guide', NRL Tech. Rpt., Naval Research Laboratory, Washington, DC, 120 p.

Imaoka K, Kachi M, Fujii H, Marakami H, Hori M, Ono A, Igarashi T, Nakagawa K, Oki T, Hoda Y & Shimoda H 2010, 'Global change observation mission (GCOM) for monitoring carbon, water cycles and climate change', *Proceedings of IEEE*, vol. 98, pp. 717-734.

Jackson TJ 1993, 'Measuring surface soil moisture using passive microwave remote sensing', *Hydrology Processes*, vol. 7, pp. 139-152.

Jackson TJ 1997, 'Soil moisture estimation using special satellite microwave/imager satellite data over a grassland region', *Water Resources Research*, vol. 33, pp.1475-1484.

Jackson TJ & Hsu AY. 2001, 'Soil moisture and TRMM microwave imager relationships in the Southern Great Plains 1999 (SGP99) experiment', *IEEE Transactions on Geoscience and Remote Sensing*, vol. 39, pp. 1632-1642

Jackson TJ & Oneill PE 1990, 'Attenuation of Soil Microwave Emission by Corn and Soybeans at 1.4 Ghz and 5 Ghz', *IEEE Transactions on Geoscience and Remote Sensing*, vol. 28, pp. 978-980.

Jackson TJ & Schmugge TJ 1991, 'Vegetation effects on the microwave emission of soils', *Remote Sensing of Environment*, vol. 36, pp. 203-212.

Kalnay E, Kanamitsu M & Baker WE 1990, 'Global numerical weather prediction at the National Meteorological Center', *Bulletin of the American Meteorological Society*, vol. 71, pp. 1410-1428.

Kalnay E, Kanamitsu M, Kistler R, Collins W, Deaven,D, Gandin L, Iredell M, Saha S, White G, Woollen J, Zhu Y, Chelliah M, Ebisuzaki W, Higgins W, Janowiak J, Mo KC, Ropelewski C, Wang J, Leetmaa A, Reynolds R, Jenne R & Joseph D 1996, 'The NCEP/NCAR 40-year reanalysis project', *Bulletin of the American Meteorological Society*, vol. 77, pp. 437-471.

Kanamitsu M 1989, 'Description of the NMC global data assimilation and forecast system', *Weather and Forecasting*, vol. 4, pp. 335-342.

Kanamitsu M, Alpert JC, Campana KA, Caplan PM, Deaven DG, Iredell M, Katz B, Pan HL, Sela J & White GH 1991, 'Recent changes implemented into the global forecast system at NMC', *Weather and Forecasting*, vol. 6, pp. 425-435.

Kerr YH, Waldteufel P, Wigneron JP, et al. 2001, 'Soil moisture retrieval from space: the soil moisture and ocean salinity (SMOS) mission. *IEEE Transactions on Geoscience and Remote Sensing*, vol. 39, pp. 1729-1735.

Koike T, Tsukamoto T, Kumakura T & Lu M 1996, 'Spatial and seasonal distribution of surface wetness derived from satellite data', *Proceedings of International workshop on macro-scale hydrological modeling*, pp. 87-96.

Li H, Robock A & Wild M 2007, 'Evaluation of Intergovernmental Panel on Climate Change Fourth Assessment soil moisture simulations for the second half of the twentieth century', *Journal of Geophysical Research*, vol. 112, D06106, doi:10.1029/2006JD007455

Liu, G 1998, 'A fast and accurate model for microwave radiance calculations', *Journal of Meteorology Society of Japan*, vol. 76, pp. 335–343.

Lu H, Koike T, Fujii H, Ohta T & Tamagawa K 2009, 'Development of a Physically-based Soil Moisture Retrieval Algorithm for Spaceborne Passive Microwave Radiometers and its Application to AMSR-E', *Journal of The Remote Sensing Society of Japan*, vol. 29, pp. 253-261.

Melillo JM, Steudler PA, Aber JD, Newkirk K, Lux H, Bowles FP, Catricala C, Magill A, Ahrens T & Morrisseau S 2002, 'Soil warming and carbon-cycle feedbacks to the climate system, *Science*, vol. 298, pp. 2173-2176.

Mo T & Schmugge TJ 1987, 'A Parameterization of the Effect of Surface-Roughness on Microwave Emission', *IEEE Transactions on Geoscience and Remote Sensing*, vol. 25, pp. 481-486

Njoku EG & Chan SK 2006, 'Vegetation and surface roughness effects on AMSR-E land observations', *Remote Sensing of Environment*, vol. 100, pp. 190-199.

Njoku EG & Entekhabi D.1996, 'Passive microwave remote sensing of soil moisture', *Journal of Hydrology*, vol.184, pp.101-129.

Njoku EG, Jackson TJ, Lakshmi V, Chan TK & Nghiem SV 2003, 'Soil moisture retrieval from AMSR-E', *IEEE Transactions on Geoscience and Remote Sensing, vol.* 41, pp. 215-229.

Njoku, EG & Li L 1999, 'Retrieval of land surface parameters using passive microwave measurements at 6-18 GHz', *IEEE Transactions on Geoscience and Remote Sensing*, vol. 37, pp. 79-93.

Oh Y & Kay YC 1998, 'Condition for precise measurement of soil surface roughness', *IEEE Transactions on Geoscience and Remote Sensing*, vol. 36, pp. 691-695.

Oh Y, Sarabandi K & Ulaby FT 1992, 'An emperical-model and an inversion technique for Radar scattering from bare soil surfaces', *IEEE Transactions on Geoscience and Remote Sensing*, vol. 30, pp.370-381.

Paloscia S, Macelloni G & Santi E 2006, 'Soil moisture estimates from AMSR-E brightness temperatures by using a dual-frequency algorithm', *IEEE Transactions on Geoscience and Remote Sensing*, vol. 44, pp. 3135-3144.

Paloscia S, Macelloni G., Santi E & Koike T. 2001, 'A multifrequency algorithm for the retrieval of soil moisture on a large scale using microwave data from SMMR and SSM/I satellites', *IEEE Transactions on Geoscience and Remote Sensing*, vol. 39, pp.1655-1661.

Paloscia S & Pampaloni P 1988, 'Microwave polarization index for monitoring vegetation growth', *IEEE Transactions on Geoscience and Remote Sensing*, vol. 26, pp. 617-621.

Paloscia, S., Pampaloni, P., Pettinato, S. & Santi, E. 2008, 'A comparison of algorithms for retrieving soil moisture from ENVISAT/ASAR images', *IEEE Transactions on Geoscience and Remote Sensing*, vol. 46, pp. 3274-3284.

Pastor J & Post WM 1986, 'Influence of climate, soil-moisture, and succession on forest carbon and nitrogen cycles', *Biogeochemistry*, vol. 2, pp. 3-27.

Prigent C, Aires F, Rossow WB & Robock A 2005, 'Sensitivity of satellite microwave and infrared observations to soil moisture at a global scale: Relationship of satellite observations to in situ soil moisture measurements', *Journal of Geophysical Research*, vol. 110, D07110. doi:10.1029/2004JD005087

Prigent C, Wigneron JP, Rossow WB & Pardo-Carrion JR 2000, 'Frequency and angular variations of land surface microwave emissivities: Can we estimate SSM/T and AMSU emissivities from SSM/I emissivities?', *IEEE Transactions on Geoscience and Remote Sensing*, vol. 38, pp.2373-2386.

Ray, PS 1972, 'Broadband Complex Refractive Indices of Ice and Water', *Applied Optics* vol. 11, pp.1836-1844.

Reichle RH & Koster RD 2005, 'Global assimilation of satellite surface soil moisture retrievals into the NASA catchment land surface model', *Geophysics research letters*, vol. 32, doi:10.1029/2004GL021700.

Schar C, Luthi D, Beyerle U & Heise E 1999, 'The soil-precipitation feedback: A process study with a regional climate model', *Journal of Climate*, vol. 12, pp.722–741.

Schmugge TJ & Jackson TJ 1992, 'A Dielectric Model of the Vegetation Effects on the Microwave Emission from Soils', *IEEE Transactions on Geoscience and Remote Sensing*, vol. 30, pp. 757-760.

Schmugge, TJ & Choudhury BJ 1981, 'A comparison of radiative transfer models for predicting the microwave emission from soils', *Radio Science*, vol. 16, pp. 927-938.

Seneviratne, SI, Luthi D, Litschi M & Schar C 2006, 'Land–atmosphere coupling and climate change in Europe', *Nature*, vol. 443, pp. 205–209.

Shi JC, Jiang LM, Zhang LX, Chen KS, Wigneron JP & Chanzy A 2005, 'A parameterized multifrequency-polarization surface emission model', *IEEE Transactions on Geoscience and Remote Sensing*, vol. 43, pp. 2831-2841.

Singh D, Sing KP, Herlin I & Sharma SK 2003, 'Ground-based scatterometer measurements of periodic surface roughness and correlation length for remote sensing', *Advances in Space Research*, vol. 32, pp. 2281-2286.

Tsang L & Kong JA 1977, 'Theory for thermal microwave emission from a bounded medium containing spherical scatterers', *Journal of Applied Physics*, vol. 48, pp. 3593-3599.

Tsang L & Kong JA 2001, *Scattering of Electomagnetic Waves: Advanced Topics*, Wiley, New York.

Uppala SM, Kållberg PW, Simmons AJ, Andrae U, V. da Costa Bechtold & Fiorino M 2005, 'The ERA-40 re-analysis', *Quarterly Journal of the Royal Meteorological Society*. Vol. 131, pp.2961–3012.

Verstraeten WW, Veroustraete F, van der Sande CJ, Grootaers I & Feyen J 2006, 'Soil moisture retrieval using thermal inertia, determined with visible and thermal spaceborne data, validated for European forests', *Remote Sensing of Environment*, vol. 101, pp. 299-314.

Wagner W, Scipal K, Pathe C, Gerten D, Lucht W & Rudolf B 2003, 'Evaluation of the agreement between the first global remotely sensed soil moisture data with model and precipitation data', *Journal of Geophysical Research*, vol. 108, pp. 4611- , doi:10.1029/2003JD003663

Wang JR & Choudhury BJ 1981, 'Remote-Sensing of Soil-Moisture Content over Bare Field at 1.4 Ghz Frequency', *Journal of Geophysical Research-Oceans and Atmospheres*, vol. 86, pp. 5277-5282.

Wang, J.R. & Schmugge, T.J. 1980, 'An empirical model for the complex dielectric permittivity of soils as a function of water content', *IEEE Transactions on Geoscience and Remote Sensing*, vol. 18, pp. 288-295.

Wegmuller U & Matzler, C 1999, 'Rough bare soil reflectivity model', *IEEE Transactions on Geoscience and Remote Sensing*, vol. 37, pp. 1391-1395.

Wen B, Tsang L, Winebrenner DP & shimura A 1990, 'Dense media radiative transfer theory: comparison with experiment and application to microwave remote sensing and polarimetry', *IEEE Transactions on Geoscience and Remote Sensing*, vol. 28, pp. 46-59.

Wigneron J. P., Schmugge T., Chanzy A., Calvet J. C. & Kerr Y. 1998, 'Use of passive microwave remote sensing to monitor soil moisture', *Agronomie*, vol.18, pp. 27-43.

Wigneron J.P., Laguerre L & Kerr Y 2001, 'A simple parameterization of the L-band microwave emission from rough agricultural soils', *IEEE Transactions on Geoscience and Remote Sensing*, vol. 39, pp. 1697-1707.

Chapter 4

Potential Impacts of and Adaptation to Future Climate Change for Crop Farms: A Case Study of Flathead Valley, Montana

Tony Prato and Zeyuan Qiu

Additional information is available at the end of the chapter

1. Introduction

Greenhouse gas emissions alter carbon and hydrologic cycles, mean surface air temperature, the spatial and temporal distribution of energy, water, and nutrients, atmospheric CO_2 concentration, and the frequency and severity of storms (Adams et al., 1990; National Research Council, 2001; Reilly 2002; Wang and Schimel 2003; Smith 2004; IPCC, 2007). A major consequence of increasing greenhouse gas emissions is climate change and variability (CCV). CCV alters annual levels and intra-annual patterns of temperature, precipitation, and other climate-related variables, which can impact crop yields and the profitability of crop farming. Such impacts are likely to vary across agricultural production areas. Crop yields are projected to increase in agricultural production areas experiencing slightly higher surface air temperature and growing season precipitation, and decrease in production areas experiencing significantly higher surface air temperature, lower growing season precipitation, and inadequate irrigation water supplies (McCarthy et al., 2001). Even if future CCV causes crop yields to decrease, crop farmers may be able to reduce those negative impacts by adapting their crop enterprises and crop production systems (CPSs) (i.e., combinations of crop enterprises) to actual or expected CCV (Stewart et al., 1998; Smit et al., 2000; Walther et al., 2002; Spittlehouse & Stewart, 2003; Antle et al., 2004; Easterling et al., 2004; Inkley et al., 2004). Most previous studies of CCV impacts on agriculture: (1) focus on how CCV is likely to impact regional or national crop yields; (2) do not consider CCV impacts on net farm income; and (3) do not evaluate the extent to which adapting crop enterprises and farms to CCV reduces adverse impacts of CCV. Because crop farming is a business, crop farmers need to consider the potential impacts of CCV on their financial returns; particularly impacts on crop enterprise net returns and net farm income.

2. Objectives

The objectives of this chapter are: (1) to assess the impacts of climate change on the levels of crop enterprise net returns and net farm income (NFI) in a future period (2006–2050) relative to their levels in an historical period (1960–2005) for small and large representative farms in Flathead Valley, Montana-the study area; and (2) to determine whether adapting CPSs to future climate change in Flathead Valley results in superior or inferior levels of net farm income compared to not adapting to future climate change. Small and large representative farms use a mix of crop enterprises, farming operations, and crop acreages, and have total sizes similar to actual small and large farms in the study area.

3. Previous research

Several studies have examined how climate change might affect agriculture. Reilly (2002) used the Hadley Center and Canadian climate models to estimate potential impacts of climate change on 2030-2090 crop yields for the entire US. He found that future climate change could result in: (1) higher yields for cotton, corn for grain and silage, soybeans, sorghum, barley, sugar beets, and citrus fruits; (2) higher or lower yields for wheat, rice, oats, hay, sugarcane, potatoes, and tomatoes, depending on the climate scenario; (3) large increases in average grain yields for the northern half of the Midwest, West, and Pacific Northwest; (4) depending on the climate scenario and time period, either increases or decreases in crop yields in other regions of the US; and (5) large reductions in crop yields in the South and Plains States for climate scenarios with low precipitation and substantial warming. For the Midwestern United States, Brown and Rosenberg (1997) simulated the impacts of climate change on crop yields and water use under different future climate scenarios using the Environmental/Policy Integrated Climate (EPIC) model (Williams et al., 1989). In a similar study, Izaurralde et al. (2003) used the EPIC model to evaluate the potential impacts of climate change on US crop yields, yield variability, incidence of various crop stress factors, evapotranspiration, and national crop production. That study evaluated how a baseline climate scenario for the period 1961–1990 and two Hadley Center climate scenarios for the periods 2025–2034 and 2090–2099 impact 204 representative farms. Reilly (2002), Brown and Rosenberg (1997), and Izaurralde et al. (2003) did not evaluate how future climate change is likely to impact crop enterprise net returns and NFI for representative farms as does this study. Kaiser et al. (1993) evaluated the economic and agronomic impacts of several climate warming scenarios, mainly temperature changes, on a grain farm in southern Minnesota and alternative ways to adapt the farm to those scenarios. That study did not evaluate the impacts of other climate variables, such as precipitation and atmospheric CO_2 concentration, on crop yields as does this study.

Antle et al. (1999) evaluated the impacts of climate change on crop enterprise returns in the Great Plains. That study showed: (1) with adaptation of crop enterprises to climate change, climate change and CO_2 enrichment caused mean crop enterprise return to change by -11% to +6% and variability in crop enterprise return to increase 7–25% relative to the baseline climate; and (2) without adaptation, mean crop enterprise return decreases 8–31% and variability in crop enterprise return increases 25–83% relative to the baseline climate. Antle et al. (2004) examined relative and absolute economic measures of the vulnerability of dryland grain farms in Montana to climate change with and without adaptation using data from a statistically representative sample of farm fields. That study allowed inferences to be drawn about the vulnerability of a heterogeneous population of farms to climate change with and without adaptation, and showed that when both climate change and higher atmospheric CO_2 concentrations are taken into account, average crop enterprise return was higher relative to the baseline climate for five and lower for three of the eight adaptation scenarios evaluated. Although Antle et al. (1999, 2004) evaluated the potential impacts of climate change on crop yields and crop enterprise returns, they did not consider potential impacts of future climate change on NFI as does this study.

4. Study area

Flathead Valley, Montana is the study area (Fig. 1). It is located in Flathead County. The county is approximately 13,605 km^2 in area (roughly the size of the State of Connecticut in the US) of which approximately 79% is managed by the federal government (Flathead County Planning and Zoning Office, 2009). In 2007, Flathead County had 1,094 farms with an average farm size of 93 ha. Of the 1,094 farms, 1,048 were less than 405 ha and 46 exceeded 405 ha in size. Sixty-one farms had annual sales less than $100,000 and 1,033 farms had annual sales greater than $100,000. In 2007, major crops grown in Flathead County in order of area harvested were spring wheat, alfalfa hay, winter wheat, other hay, barley, and canola. In 2006, cash receipts from the sale of principal agricultural commodities in Flathead County amounted to $33.5 million (Montana Agricultural Statistics Service, 2008; National Agricultural Statistics Service, 2011).

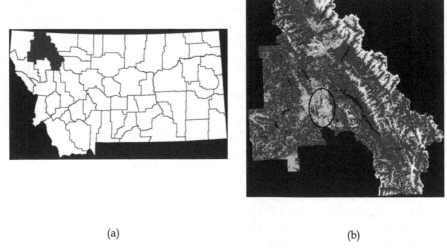

(a)   (b)

Figure 1. Location of Flathead County in Montana (a) and Flathead Valley (oval-shaped area) (b)

Daily and monthly climate data from the Creston, Montana weather station located in the Flathead Valley show that during the historical period: (1) average monthly maximum and average monthly minimum temperature for the winter months (i.e., December through February) were 0.41°C and -7.93°C, respectively; (2) average monthly maximum temperature and average monthly minimum temperature for the summer months (i.e., June through August) were 24.88°C and 8.32°C, respectively; and (3) average annual precipitation was 488 mm.

5. Methods and procedures

This section begins with an overview of the methods and procedures used to assess the potential agricultural impacts of the three climate scenarios (i.e., impact assessment) and the potential benefits of adapting CPSs for representative farms to the three climate scenarios (i.e., adaptation evaluation), and describes in detail the methods and procedures used in the impact assessment and adaptation evaluation.

5.1. Overview

The impact assessment determines the potential agricultural impacts of CO_2 emissions scenarios A1B, B1, and A2 developed by the Intergovernmental Panel on Climate Change Fourth Assessment Report (IPCC) (2007). The assessment involved: (1) specifying crop enterprises, CPSs, and soil types for small and large representative farms in Flathead Valley; (2) simulating crop yields; and (3) estimating net returns for crop enterprises and CPSs and net farm income in the historical and future periods. The adaptation evaluation determined the potential benefits of adapting CPSs for representative farms to the three climate scenarios, which involved determining: (1) whether the most dominant CPS in the historical period is different than the most dominant CPS for each climate scenario in the future period; and (2) for cases where they are different, whether the most dominant CPS for each climate scenario is superior to the most dominant CPS in the historical period.

5.2. Impact assessment

5.2.1. Specifying crop enterprises, CPSs, soils types, and representative farms

Ten crop enterprises, common to Flathead Valley, were specified for the study: spring wheat; winter wheat; oats; spring canola; spring barley; dryland (unirrigated) alfalfa; irrigated alfalfa; spring lentils; and dry (unirrigated) peas. Permanent pasture, a common forage crop enterprise in Flathead Valley, was excluded from the study because it does not produce a marketed crop. A CPS is a unique combination of crop enterprises. Two producer panels were established; one for a small representative farm (66 ha) and the other for a large representative farm (243 ha). Each panel consisted of 3-5 farmers that operated a small-scale or large-scale farm in Flathead Valley. Three CPSs were specified for each representative farm (Table 1) with the assistance of the producer panels. Two common soil types were evaluated for each crop enterprise: Creston silt loam (Ce), which is on 0–3% slopes and accounts for 3.4% of the total agricultural area in Flathead Valley; and Kalispell loam (Ke), which is on 0–3% slopes and accounts for 2.7% of the total agricultural area in Flathead Valley.

Crop enterprise	Large representative farm		
	CPS 1	CPS 2	CPS 3
Spring wheat	–	81	162
Winter wheat	81	–	–
Oats	–	–	40
Spring canola	40	–	–
Spring barley	61	61	–
Dry alfalfa	61	–	–
Irrigated alfalfa	–	61	–
Spring lentils	–	–	40
Dry peas	–	40	–
	Small representative farm		
	CPS 4	CPS 5	CPS 6
Spring wheat	–	12	8
Oats	–	8	–
Spring canola	12	8	8
Spring barley	10	–	–
Irrigated alfalfa	28	22	22
Spring lentils	–	–	12

Table 1. Hectares in crop enterprises for crop production systems (CPSs) for large and small representative farms

5.2.2. Simulating crop yields

Annual crop yields in the historical period were simulated for both soil types by inputting to the EPIC model (Williams et al., 1989) daily data on precipitation, maximum temperature (T_{max}), minimum temperature (T_{min}), relative humidity, solar radiation, and wind velocity from the Creston weather station in Flathead Valley and other sources, and field operations for crop enterprises (i.e., amount and/or timing of planting, fertilizer/pesticide use, tillage operations, and harvesting). Annual atmospheric CO_2 concentrations for the historical period were determined using the dynamic CO_2 option in the EPIC model. That option varies the annual atmospheric CO_2 concentration according to the following quadratic equation: $CO_2 (X) = 280.33 - 0.1879X + 0.0077X^2$; where X equals the number of years between the prediction year and 1880. For example, for X = 2000 - 1880 = 120, the equation gives a CO_2 concentration in 2000 of $CO_2 (120) = 280.33 - 0.1879 * 120 + 0.0077 * (120)^2 = 368.7$ ppm. This regression equation was estimated using the historical CO_2 record from the Mauna Loa Observatory in Hawaii (Izaurralde et al., 2006).

Annual crop yields in the future period were simulated for each climate scenario and soil type by inputting to the EPIC model daily projections of precipitation, maximum temperature (T_{max}), minimum temperature (T_{min}), relative humidity, solar radiation, and wind velocity, and annual projections of atmospheric CO_2 concentration for that climate scenario. Daily projections of precipitation and temperature were derived by applying the delta method (e.g., McGinn et al., 1999) to monthly bias-corrected, downscaled climate projections for each of the three climate scenarios. Monthly climate projections are based on the World Climate Research Program's (WCRP's) Coupled Model Intercomparison Project phase 3 (CMIP3) (Meehl et al., 2007), which are available through the Program for Climate Model Diagnosis and Intercomparison (PCMDI) and the WCRP's Working Group on Coupled Modeling (WGCM; see http://gdo-dcp.ucllnl.org/). CMIP3 climate projections synthesize monthly temperature and precipitation data from 112 projection-specific datasets representing 16 CMIP3 climate models and the three future CO_2 emission scenarios for the period 1950-2099 (Meehl et al., 2007). In terms of which downscaled climate projection to use, the 12-km downscaled grid used in the study is the grid centered over the Creston meteorological station and the majority of the Flathead Valley. Annual CO_2 concentrations for each climate scenario were interpolated assuming linear increases in CO_2 concentration from 379 ppm in 2005 to the IPCC concentration for that scenario specified in 2100 (Table 2). The crop yield simulations take into account the fertilization effects of CO_2 concentration for the three climate scenarios.

The CMIP3 dataset does not contain monthly data on relative humidity, solar radiation, and wind velocity. Daily relative humidity and solar radiation projections were developed by applying the MTCLIM model (Hungerford et al., 1989; Kimball et al., 1997) to the daily temperature and precipitation projections for each climate scenario. Due to lack of data, daily wind velocity for each climate scenario was assumed to be the same as the corresponding daily wind velocity in the historical period. Specifically, daily wind velocity in month t+30 for each climate scenario equals the corresponding daily wind velocity in month t of the historical period.

EPIC simulates annual crop yields for the 46 years in the historical period and 45 years in the future period based on operations for each crop (i.e., amount and/or timing of planting, fertilizer/pesticide use, tillage operations, and harvesting) specified by producer panels, weather data for the historical period, and weather projections for each climate scenario in the future period discussed above. These simulated annual crop yields are referred to as raw crop yields, which are then used to extract the underlying crop yield distribution and derive 100 values of crop yields for calculating crop enterprise net returns. The parameter estimation, simulate, and CDFDEV functions in the Simulation and Econometrics to Analyze Risk (Simetar) program (Richardson et al., 2006) were used to simulate 100 values of crop yields for each period as follows. First, the parameter estimation function with maximum likelihood estimation was used to fit 16 probability distributions (i.e., Beta, double exponential, exponential, gamma, logistic, log–log, log–logistic, lognormal, normal, Pareto, uniform, Weibull, binomial, geometric, Poisson, and negative binomial) to raw crop yields for the historical period and each climate scenario in the future period. Second, the simulate function in Simetar was applied to the estimated parameters of each fitted probability distribution to simulate 100 values of crop yields for each distribution. Third, the CDFDEV function was applied to the 100 simulated crop yields for each distribution to determine the best-fitting probability distribution for crop yields. Fourth, a random sample of 100 values of crop yields was drawn from the best-fitting probability distribution was used to calculate 100 values of net returns per ha.

Scenario	Level of forcing	CO_2 concentration in 2100 (ppm)	Average increase in global temperature (°C)[a]
B1	Low	530	1.8
A1B	Medium	700	2.8
A2	High	800	3.4

[a]Mean temperature for years 2090-2099 minus mean temperature for years 1980-1989, Source: IPCC (2007)

Table 2. Description of three climate scenarios

5.2.3. Estimating net returns for crop enterprise and CPSs and net farm income

Annual net return per ha for a crop enterprise equals annual crop yield times crop price per unit of output minus total cost of production per ha. The 100 values of crop yields simulated using the procedures described in section 5.2.2, and 100 values of crop prices per unit of output and annual total cost of production per ha were used to simulate 100 values of crop enterprise net returns for the historical period and each climate scenario for the future period. The 100 values of crop prices were randomly selected from the best-fitting probability distribution for crop prices determined by applying the three Simetar functions described in section 5.2.2 to annual inflation-adjusted (base year = 2008) crop prices for the historical climate period. The 100 values of total cost of production per ha for a crop enterprise were randomly selected from triangular probability distributions. The mean of the triangular distribution equals the mean annual total cost of production per ha for that crop enterprise given by crop enterprise budgets for the study area (Table 3). Mean annual total cost of production is the sum of average variable and average fixed costs per ha. Variable cost includes the costs of seed, fertilizer, pesticides, fuel and lubricants, hired labor, and, in the case of irrigated crops, the cost of pumping and applying irrigation water. Fixed cost includes the costs of land, equipment, machinery, vehicles, and owner/operator labor. The minimum value of the triangular probability distribution was set equal to 80% of the mean and the maximum value was set equal to 120% of the mean. It was assumed that inflation-adjusted crop prices per unit of output and mean annual total cost of production per ha in the future period were the same as in the historical period. For that reason, the same 100 values of crop prices and total cost of production per ha randomly selected for a given crop enterprise for the historical period were used for the future period.

5.3. Adaptation evaluation

The dominant CPS was identified for each representative farm and period by applying the stochastic efficiency with respect to a function (SERF) criterion (Hardaker et al., 2004) for a particular risk aversion coefficient (RAC) to the 100 simulated values of net returns for the three

CPSs specified for each farm. With the SERF criterion, the dominant CPS for a representative farm is the one with the highest certainty equivalent (Hardaker et al., 2004). The latter is the payoff amount that a farmer is willing to receive in exchange for accepting the variability in NFI associated with a particular CPS. Application of the SERF criterion was based on three assumptions: (1) RACs are in the range [0,0.03], where RAC = 0 implies the farmer is risk neutral and RAC > 0 implies the farmer is risk averse (Anderson and Dillon, 1992); (2) constant absolute risk aversion, which implies that the risk premium a farmer is willing to pay to reduce income risk does not vary with the level of income; and (3) the farmer's utility function is exponential in NFI (i.e., u[NFI] = exp[-RAC * NFI]). In addition, the SERF criterion was used to determine whether the dominant CPS for a representative farm in the historical period is superior to the dominant CPS for that representative farm under each climate scenario. If the dominant CPS in the historical period is the same as the dominant CPS under a climate scenario in the future period, then adaptation to that climate scenario is not advantageous. If the dominant CPS in the historical period differs from the dominant CPS under a climate scenario (e.g., CPS_i is the dominant CPS in the historical period and CPS_j is the dominant CPS for climate scenario k) and CPS_j dominates CPS_i, then adapting CPSs to that climate scenario (i.e., switching from CPS_i to CPS_j under climate scenario k), is advantageous to the farmer.

Crop enterprise	Variable cost	Fixed cost	Total cost[d]
Spring wheat	307.54[a]	91.07[a]	398.61
Winter wheat	274.42[b]	117.74[b]	392.16
Oats	227.51[b]	129.03[b]	356.54
Spring canola	383.59[a]	57.65[a]	441.24
Spring barley	307.74[a]	91.07[ac]	398.81
Dry alfalfa	152.23[a]	95.96[a]	248.19
Irrigated alfalfa	498.67[a]	95.96[ab]	594.63
Spring lentils	247.22[b]	130.47[b]	377.69
Dry peas	253.67[b]	130.47[b]	384.13

[a]Based on crop enterprise budgets supplied by Duane Johnson, former Superintendant of Montana State University's Northwestern Montana Agricultural Research Center, Creston, MT
[b]Based on predicted 2008 crop enterprise budgets for northwest North Dakota

Table 3. Variable, fixed, and total cost for crop enterprises ($per ha in 2008 dollars)

6. Results

6.1. Impact assessment

Simulated annual crop yields for the same crop were very similar across the three climate scenarios because IPCC climate projections of monthly temperature and precipitation are very similar across the three climate scenarios. The latter occurs because the divergence in the time paths of temperature and precipitation for the three climate scenarios does not take place until the latter half of the IPCC assessment period (i.e., 2055–2100), which occurs after the future period. Because simulated yields for a given crop are very similar across the three climate scenarios and the 100 simulated crop prices and production costs for a given crop are the same across the three scenarios, crop enterprise net returns for the same crop and soil type and NFI for the same CPS and soil type are likewise similar across the three climate scenarios. For that reason, results for the future period are averages of the results for the three climate scenarios.

6.1.1. Means and standard deviations of simulated crop enterprise net returns

Means and one-standard deviation error bars for simulated crop enterprise net returns per ha are given in Fig. 2 for the historical period and Fig. 3 for the future period. Between the historical and future periods, enterprise net returns: (1) decreases by 84.3% on average for spring barley, dry canola, dry and irrigated alfalfa, oats (in Ce soil), and spring wheat (in Ce soil); and (2) increases by 44% on average for dry lentils, oats (in Ke soil), winter wheat, spring wheat (in Ke soil), and dry peas. Averaged over the nine crop enterprises and two soil types, mean simulated crop enterprise net return per ha is 24% lower in the future period than in the historical period. In summary, mean simulated net return per ha for the same crop enterprise is between 202% lower and 74% higher in the future period than in the historical period.

6.1.2. Means and standard deviations of net farm income for crop production systems

The mean and one-standard deviation error bars for simulated NFIs for the six CPSs in the historical period are shown in Fig. 4 for the historical period and Fig. 5 for the future period. Simulated NFIs for CPSs in the future period assume no adaptation to climate scenarios. As expected, for both periods, simulated NFI is higher for the large representative farm (i.e., CPS 1, CPS 2, and CPS 3) than for the small representative farm (i.e., CPS 4, CPS 5, and CPS 6). In four of the six cases in the historical period, mean simulated NFI is higher for Ke soil than Ce soil. For the historical period and large representative farm, the mean simulated NFI is highest for CPS 3 in Ce soil at $87,275 and lowest for CPS 1 in Ce soil at $65,568. For the historical period and small representative farm, mean simulated NFI is highest for CPS 6 in Ke soil at $23,612 and lowest for CPS 4 in Ce soil at $21,599. For the future period and large representative farm, the mean simulated NFI is highest for CPS 3 in Ke soil for climate scenario B1 at $40,571 and lowest for CPS 3 in Ce soil for climate scenario A2 at $14,585. For the future period and small representative farm, the mean simulated NFI is highest for CPS 6 in Ke soil for climate scenario A2 at $13,726 and lowest for CPS 5 in Ce soil for climate scenario A2 at $8,864. Mean simulated NFIs for the CPSs decrease 57% between the historical and future periods. The maximum percent decline in mean simulated NFI between the historical and future periods is 83% for CPS 3 in Ce soil for the large representative farm under climate scenario A2. The minimum percent decline in mean simulated NFI between the historical and future periods is 41.9% for CPS 6 in Ke soil for the small representative farm under climate scenario A2. In summary, mean simulated net farm income for the same CPS is between 42% and 83% lower in the future period than in the historical period.

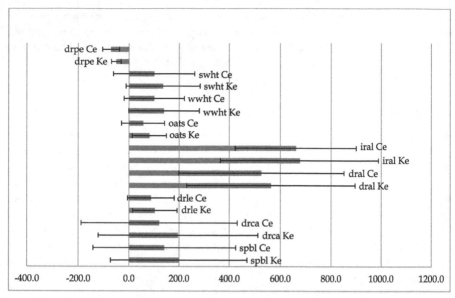

Figure 2. Means and one-standard deviation error bars for simulated crop enterprise net returns per ha (in 2008 dollars) for the historical period, by soil type (drpe is dry peas, swht is spring wheat, wwht is winter wheat, iral is irrigated alfalfa, dral is dry (unirrigated) alfalfa, sple is spring lentils, drca is dry (unirrigated) canola, and spbl is spring barley)

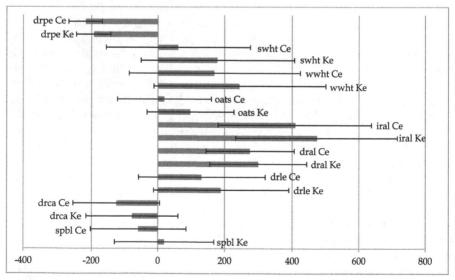

Figure 3. Means and one-standard deviation error bars for simulated crop enterprise net returns per ha (in 2008 dollars) for the future period, by soil type (crop enterprise legend given in Fig. 2)

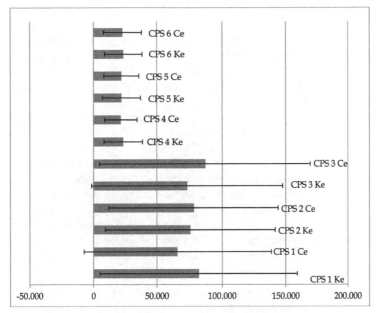

Figure 4. Means and one-standard deviation error bars for simulated net farm income in 2008 dollars for crop production systems (CPSs) in the historical period, by soil type (crop enterprises for each CPS are listed in Table 1)

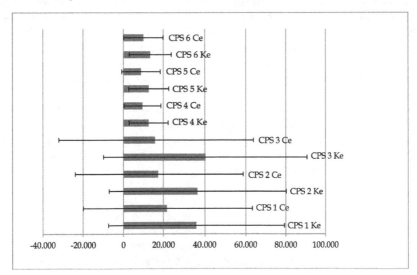

Figure 5. Means and one-standard deviation errors bars for simulated net farm income in 2008 dollars for crop production systems (CPSs) for the future period without adaptation to climate scenarios, by soil type (crop enterprises for each CPS are listed in Table 1)

6.2. Adaptation evaluation

Table 4 shows the dominant CPSs for the small and large representative farms and two soil types for the historical period and three climate scenarios. For the large representative farm: (1) CPS 2 dominates CPS 1 and CPS 3 for both soil types in the historical period and under climate scenarios A1B and A2 for the Ke soil type; (2) CPS 3 dominates CPS 1 and CPS 2 under climate scenario B1 for the Ke soil type; and (3) CPS 1 dominates CPS 2 and CPS 3 under all three climate scenarios for the Ce soil type. For the small representative farm: (1) CPS 4 dominates CPS 5 and CPS 6 in the historical period for both soil types; (2) CPS 4 dominates CPS 5 and CPS 6 under climate scenario A1B for both soil types; and (3) CPS 5 dominates CPS 4 and CPS 6 under climate scenarios A2 and B1 for both soil types. These results indicate that switching CPSs between the historical and future periods (i.e., adapting CPSs to future climate change) is optimal in eight of the twelve cases evaluated. Specifically, it is advantageous to switch: (1) from CPS 2 to CPS 3 under climate scenario B1 for the Ke soil type and from CPS 2 to CPS 1 under all three climate scenarios for the Ce soil type for the large representative farm; and (2) from CPS 4 to CPS 5 under climate scenarios A2 and B1 for both soil types for the small representative farm.

Table 5 reports the dominance relationships for CPSs for the large and small representative farms and two soil types between the historical period and three climate scenarios. Of particular interest are the dominance relationships for the cases in which the dominant CPS differs between the historical and future periods because these relationships indicate whether NFI in the historical period is superior or inferior to NFI with adaptation of CPSs to climate change. For the large representative farm: (1) CPS 2 in the historical period dominates CPS 3 under climate scenario B1 for soil type Ke; and (2) CPS 2 in the historical period dominates CPS 1 under all three climate scenarios for the Ce soil type. For the small representative farm: (1) CPS 4 in the historical period dominates CPS 5 under climate scenarios B1 and A2 for soil type Ke; and (2) CPS 5 under climate scenarios B1 and A2 dominates CPS 4 in the historical period for the Ce soil type. Combining the results in Tables 4 and 5 indicates that while adapting CPSs to future climate change reduces potential losses in NFI in eight of the 12 cases evaluated, in only three of those eight cases is NFI in the future period after adaptation to climate change superior to NFI in the historical period. Conversely, in five of those eight cases, NFI in the future period after adaptation to climate change is inferior to NFI in the historical period.

7. Conclusion

It is difficult to evaluate the potential adverse impacts of future climate change on agricultural production because of uncertainty regarding the nature and extent of future climate change and how such change is likely to influence crop yields, crop enterprise net returns, and NFIs for CPSs. Most previous studies of the agricultural impacts of climate change focus on how past climate change has influenced crop yields and/or crop enterprise net returns at the regional and/or national levels. The unique contribution of this study is that it developed a method for assessing the potential economic benefits (in terms of alleviating losses in NFI) of adapting CPSs to future climate change for representative farms in a local agricultural production area. This is an important contribution because farming is a business that requires farmers to understand the potential impacts of future climate change on NFI and determine whether adapting CPSs to future climate change alleviates negative impacts of those changes on NFI.

Averaged over the two representative farms and two soil types in Montana's Flathead Valley, simulated net return per ha for the nine crop enterprises decreases 24% and mean simulated NFI for CPSs decreases 57% between the historical and future periods. Although adapting CPSs to future climate change reduces potential losses in NFI in eight of the 12 cases evaluated here, in only three of those eight cases is NFI in the future period after adaptation to climate change superior to NFI in the historical period. Therefore, for most part, adapting CPSs to future climate change alleviates but does not eliminate the negative impacts of that change on simulated NFI. The impact assessment and adaptation evaluation methods described here can be used to determine the potential impacts of future climate change on crop enterprise net returns and NFI for representative farms and evaluate the potential economic benefits of adapting crop enterprises and CPSs to future climate change in other agricultural production areas.

Soil type	Large representative farm	Small representative farm
	Historical period	
Ke	CPS 2	CPS 4
Ce	CPS 2	CPS 4
	Climate scenario	
	A1B	
Ke	CPS 2	CPS 4
Ce	CPS 1	CPS 4
	A2	
Ke	CPS 2	CPS 5
Ce	CPS 1	CPS 5
	B1	
Ke	CPS 3	CPS 5
Ce	CPS 1	CPS 5

[a]Based on SERF method assuming a risk-averse farmer (i.e., 0.0013 < RAC ≤ 0.03)

Table 4. Dominant crop production systems (CPSs) for the historical period and each of the three climate scenarios (B1, A1B, and A2), by large and small representative farms and two soil types[a]

Soil Type	Large representative farm	Small representative farm
Ke	CPS 2 (H) $\mathcal{D}$[b] CPS 3 (B1)	CPS 4 (H) $\mathcal{D}$ CPS 5 (B1)
	CPS 2 (H) $\mathcal{D}$ CPS 2 (A1B)	CPS 4 (H) $\mathcal{D}$ CPS 4 (A1B)
	CPS 2 (H) $\mathcal{D}$ CPS 2 (A2)	CPS 4 (H) $\mathcal{D}$ CPS 5 (A2)
Ce	CPS 2 (H) $\mathcal{D}$ CPS 1 (B1)	CPS 5 (B1) $\mathcal{D}$ CPS 4 (H)
	CPS 2 (H) $\mathcal{D}$ CPS 1 (A1B)	CPS 4 (A1B) $\mathcal{D}$ CPS 4 (H)
	CPS 2 (H) $\mathcal{D}$ CPS 1 (A2)	CPS 5 (A2) $\mathcal{D}$ CPS 4 (H)

[a]Based on SERF method assuming a risk-averse farmer (i.e., $0.0013 < RAC \leq 0.03$)
[b]$\mathcal{D}$ indicates "dominates"

Table 5. Dominance relationships for crop production systems (CPSs) across the historical period (H) and three climate scenarios (B1, A1B, and A2), by large and small representative farms and two soil types[a]

Author details

Tony Prato
University of Missouri, USA

Zeyuan Qiu
New Jersey Institute of Technology, USA

Acknowledgement

The research reported here was supported in part by the National Research Initiative of the USDA Cooperative State Research, Education and Extension Service, grant number 2006-55101-17129.

8. References

Adams, R.M.; Rosenzweig, C., Peart, R.M., Ritchie, J.,T., McCarl, B.A., Glyer, J.D., Curry, R.B., Jones, J.W., Boote, K.J., & Allen, L.H., Jr. (1990). Global climate change and US agriculture. *Nature* 345:219–224

Anderson, J.R.; & Dillon, J.L. (1992). Risk analysis in dryland farming systems. Farm systems management series No. 2, Food and Agriculture Organization, Rome

Antle, J.M.; Capalbo, S.M., & Hewitt, J. (1999). Testing hypotheses in integrated impact assessments: climate variability and economic adaptation in Great Plains agriculture. National Institute for Global Environmental Change, Nebraska Earth Science Education Network, University of Nebraska, Lincoln, NE, pp T5-4 and 21

Antle, JM.; Capalbo, S., Elliott, E. & Paustian, K.H. (2004). Adaptation, spatial heterogeneity, and the vulnerability of agricultural systems to climate change and CO_2 fertilization: an integrated assessment approach. *Climatic Change* 64:289–315

Brown, R.A.; & Rosenberg, N.J. (1997). Sensitivity of crop yield and water use to change in a range of climatic factors and CO_2 concentrations: a simulation study applying EPIC to the central USA. *Agricultural and Forest Meteorology* 83:171–203

Easterling, W.E. III; Hurd, B.H., & Smith, J.B. (2004). Coping with global climate change: the role of adaptation in the United States. Pew Center on Global Climate Change, Arlington, VA

Flathead County Planning and Zoning Office. (2009). Flathead county growth policy, Chap. 2: land uses, Available from

http://flathead.mt.gov/planning_zoning/growthpolicy/Chapter%202%20April%2010.pdf

Hardaker, J.B.; & Gudbrand, L. (2003). Stochastic efficiency analysis with risk aversion bounds: A simplified approach. Graduate School of Agricultural and Resource Economics & School of Economics, University of New England, Armidale, New South Wales, Australia

Hungerford, R.D.; Nemani, R.R., Running, S.W., & Coughlan, J.C. (1989). MTCLIM A Mountain Microclimate Simulation Model, USDA Forest Service, Research Paper INT-414

Inkley, D.B.; Anderson, M.G., Blaustein, A.R., Burkett, V.R., Felzer, B., Griffith. B., Price, J. & Root, T.L. (2004). Global climate change and wildlife in North America. Technical review 04-2. The Wildlife Society, Bethesda, MD

Intergovernmental Panel on Climate Change (IPCC). 2007. Climate Change 2007: Synthesis Report. Cambridge University Press, UK, Available from

http://www.ipcc.ch/publications_and_data/publications_ipcc_fourth_assessment_report
_synthesis_report.htm

Izaurralde, R.C.; Rosenberg, N.J., Brown, R.A., Allison, J.R., Thomson, A.M. (2003). Integrated assessment of Hadley Centre (HadCM2) climate-change impacts on agricultural productivity and irrigation water supply in the conterminous United States. Part II. Regional agricultural production in 030 and 095. *Agricultural and Forest Meteorology* 117:97–122

Izaurralde, R.C.; Williams, J.R., McGill, W.B., Rosenberg, N.J., & Quiroga Jakas, M.C. (2006). Simulating soil C dynamics with EPIC: model description and testing against long-term data. *Ecological Modelling* 192:362–384

Kaiser, H.M.; Riha, S.J., Wilks, D.S., Rossiter, D.G., & Sampath, F. (1993). A farm-level analysis of economic and agronomic impacts of gradual climate warming. *American Journal of Agricultural Economics* 75:387–398

Kimball, J.S.; Running, S.W., & Nemani, R. (1997). An improved method for estimating surface humidity from daily minimum temperature. *Agricultural and Forest Meteorology* 85:87–98

McCarthy, J.J.; Canziani, O.F., Leary, N.A., Dokken, D.J. & White, K.S. (eds) (2001). Climate change 2001: impacts, adaptation, and vulnerability. Intergovernmental panel on climate change, Cambridge University Press, UK

McGinn, S.M.; Toure, A., Akinremi, O.O., Major, D.J., & Barr, A.G. (1999). Agroclimate and crop response to climate change in Alberta, Canada. *Outlook on Agriculture* 28:19–28

Meehl, G.A.; Covey, C., Delworth, T., Latif, M., McAvaney, B., Mitchell, J.F.B., Stouffer, F.J., & Taylor, K.E. (2007). The WCRP CMIP3 multimodel dataset: a new era in climate change research. *Bulletin of the American Meteorological Society* 88:1383–1394

Montana Agricultural Statistics Service. (2008). County Profiles. Flathead County, Available from

http://www.nass.usda.gov/Statistics_by_State/Montana/Publications/county/profiles/20 07/C07029.htm

National Agricultural Statistics Service. (2011). Statistics by States, Available from http://www.nass.usda.gov/Statistics_by_State/index.asp

National Research Council. (2001). Climate change science: an analysis of some key questions. Committee on the science of climate change, National Academy Press, Washington, DC

Reilly, J.M. (2002). Agriculture: the potential consequences of climate variability and change. A report of the National Agriculture Assessment Group for the U.S. Global Change Research Program. Cambridge University Press, UK

Richardson, J.W.; Schumann, K.D., & Feldman, P.A. (2006). Simetar: simulation and econometrics to analyze risk. College Station, TX, Available from http://www.simetar.com/Browser.aspx

Smit, B.; Burton, I., Klein, R. & Wandel, J. (2000). An anatomy of adaptation to climate change and variability. *Climatic Change* 45:223–251

Smith, J.B. (2004). A synthesis of potential climate change impacts on the U.S. Pew Center on Global Climate Change, Arlington, VA

Spittlehouse, D.L.; & Stewart, R.B. (2003). Adaptation to climate change in forest management. *BC Journal of Ecosystems and Management* 4:7–17, Available from http://www.forrex.org/jem/archiveVol4iss1.asp

Stewart, R.B.; Wheaton, E. & Spittlehouse, D.L. (1998). Climate change: implications for the boreal forest. In: Legge AH, Jones LL (eds) Emerging air issues for the 21st century: the need for multidisciplinary management. Air and Waste Management Association, Pittsburg, PA, pp. 86–101

Walther, G-R; Post, E., Convey, P., Menzel, A., Parmesan, C., Beebee, T.J.C., Fromentin, J-M, Hoegh-Guldberg, O., & Bairlein, F. (2002). Ecological responses to recent climate change. *Nature* 416:389–395

Wang, G.; & Schimel, D. (2003). Climate change, climate modes, and climate impacts. *Annual Review of Environment and Resources* 28:1–128

Williams, J.R.; Jones, C.A., Kiniry, J.R., & Spanel, D.A., 1989. The EPIC crop growth model. *Transactions of the ASAE* 32:497–511

Climate Change and Its Impacts on Uncertainties/Risks

Study on Perspectives of Energy Production Systems and Climate Change Risks in Nigeria

S.C. Nwanya

Additional information is available at the end of the chapter

1. Introduction

In recent times, energy has been the hottest globally discussed subject. On the other hand, environment is the most resilient victim of the energy debate. Consequently, energy and environment are the world's most unlikely allies. Energy extraction, distribution or consumption constitutes a major cause of environmental pollution. The environmental pollutants from energy related activities are greenhouse gases. Although, cement manufacturing, construction or transportation activities do contribute to environmental pollution, the greenhouse gas emission from energy activities is two-fold: the emission from exploration and that from consumption. The combustion of energy fuels generate nitrogen oxides- a group of highly reactive and acidifying gases unlike suspended particles produced from cement manufacturing. In a photochemical process, nitrogen oxides are oxidized to nitric acid and it contributes to acid rain formation. Also, there is a consensus that fossil fuel based energy production and use are the main sources of carbon dioxide and other greenhouse gas emissions as shown in figure 1. These substances have many consequences for the health of human being, plants and estate property [1, 2]. The foregoing facts present energy production and utilization as high risk factors to the environment despite their huge benefits to the society.

The generic factors that cause the emissions are classified in two ways: anthropogenic and natural occurrences. The main anthropogenic contributors are identified as follows:

- carbon emissions from industrial processes
- agriculture (methane emissions from livestock and manure, and nitrous oxide emissions from chemical fertilisers)
- carbon emissions from transport (driving a car, air travel)
- use of fuel to generate energy (excluding transport)
- energy use in the home (the main use is heating)

- deforestation

On the other hand, the natural contributors to greenhouse gas emission are:

- Volcanic eruptions
- Ocean current
- Earth orbital changes
- Solar variability
- Tectonic processes

The CO_2 emissions have through a complex relationship formed with other anthropogenic gases reduced the capacity of the atmosphere to filter out the sun's harmful ultraviolet radiations, thus causing climate change [3].

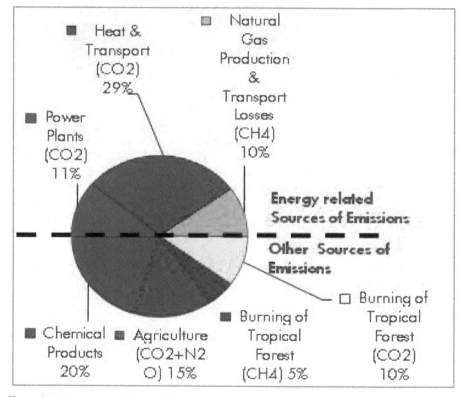

Figure 1. Major sources of the greenhouse gases.

Given the aforementioned causative factors, understanding the complex climate change, its concept or solution requires a trans-disciplinarity interpretation. Trans-disciplinarity, is the principle of integrating forms of research comprising a family of methods for relating scientific knowledge and extra-scientific experience and practice in problem-solving. It is a form of joint problem solving among science, technology and society [4]. The United Nations Framework Convention on Climate Change (UNFCCC) defines climate change as a change of climate which is attributable directly or indirectly to human activity that alters the composition of the global atmosphere and which is in addition to natural climate variability observed over a comparable time periods [5]. Different authors have attempted to define climate change according to varied disciplines to reflect region or country-specific effects.

However, the common denominator of the definitions is that climate change is an extreme effect of climatic conditions triggered off by certain actions that are either naturally occurring or having human origin. Though, climate change is a global phenomenon, country-specific effects or adversities vary over time. The impacts vary by type of causative agent and geographical location of the vulnerable country.

Developing countries like Nigeria constitute the major hotspots of the climate change. Their fast growing population, crude agricultural practices (for nitrous oxide source), spiralling energy demand and lack of care for the environment are responsible for the peculiar situation. Definitely there are implications of these to energy producing and exporting countries like Nigeria, where there are multiple sources of emissions of greenhouse gases. According to Iloeje [6], Nigeria's energy reserves constitute both an opportunity and a risk. The former relates to benefits such as huge revenue generation while the later is associated with the burden on the ecosystem such as climate change.

This study focuses on Nigeria because of rising evidences and claims about prolonged changes in the climatic conditions. Various studies have identified impacts on the environment that are linked to climate change and are mostly caused by anthropogenic factors [7, 8]. Specifically, Nworah [9] has shown evidence of climate change impacts on the environment due to increase fossil energy production and consumption in Nigeria. Also, in the energy sector gas flare is a major contributor to the air pollution [10], which results in climate change. Figures 2, 3, and 4 showed evidence of ambient temperature rise in the last 100 years in Nigeria. Therefore, climate change poses great risks to health, built-up environment and social well-being.

However, the above studies have not done significant work on the adaptation and mitigation strategies to curb the effects of climate change. Developing high capacity for climate change adaptation and CO_2 emissions mitigation strategies is a necessity and priceless tool to overcome lasting effects of the change. Dearth of these strategies has been observed more in developing countries, such as Nigeria, than the developed countries. It is for the aforementioned reason that developing countries are highly vulnerable to climate change hazards. Getting sustainable solutions for these hazardous impacts is the motivation of this study.

The premium objective of the study is to identify policy and technological innovations and best practices required to advance climate change resilience for the environment as well as to promote low carbon development in Nigeria. It is believed that by domesticating adaptation strategies Nigeria will be well of it.

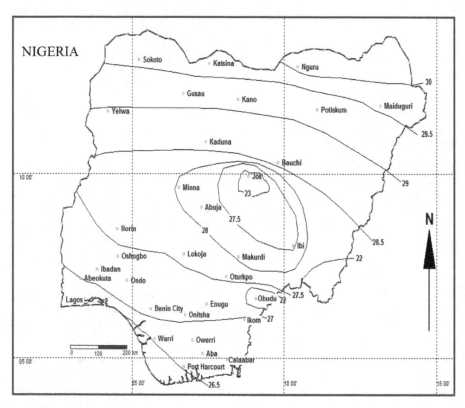

Source: [11]

Figure 2. Spatial pattern of air temperature rise in Nigeria (1901 – 1935)

2. An overview of the Nigerian energy production system in relation to climate change

The choice of Nigeria for this study is on account of the following:

Population (2006 census): 140 million (there is one Nigerian in every three Africans)

Size: 923,770 square kilometres (With density of 152 person per square kilometre climate change problem could be worse that first thought)

Location: Lies between latitudes 4^0 and 14^0 N and longitudes 3^0 and 15^0E. The location is characterised by a variety of climatic regimes such as tropical rainforest along the coasts, humid and Sahel regions in the south and north, respectively.

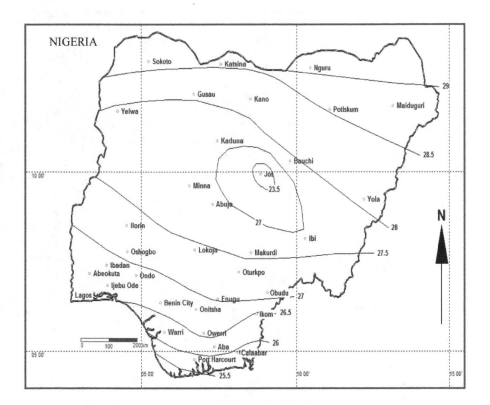

Source: [11]

Figure 3. Spatial pattern of air temperature rise in Nigeria (1936 – 1970).

Coastal risks: one –third of the 36 states – Lagos, Rivers, Ogun, Cross River, Bayelsa, Akwa Ibom, Abia, Imo, Ondo, Delta and Edo live within 10 to 80 Km of the Atlantic Ocean, a low lying region and are at risk from increased storm intensity and flooding.

Fossil reserves: 5.1 MMtcm (10^6 Trillion m^3) of proven natural gas, oil

Total energy consumption (2002E):275 billion kWh (0.2 % world total energy consumption)

Fuel share of energy consumption: Oil (58%), Natural gas (34%), Hydroelectric (7.9%), Coal (0.1%).

Energy related carbon dioxide emissions (2002E): 91.94 million metric tons (0.4% of world record).

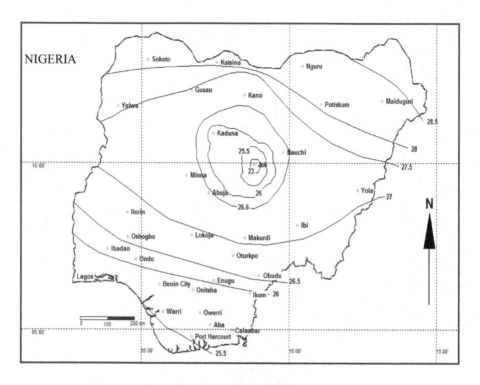

Source:[11]

Figure 4. Ambient temperature rise in Nigeria (1971- 2005)

Per capita energy consumption (2002E): 1,964.3 kWh (vs U.S. value of 99,356.3 kWh)

Per capita C02 emissions (2002E): 0.8 metric tons (vs U. S. value of 19.97 metric tons)

Energy intensity (2002E): 2.712 kWh/$ nominal-PPP (vs U. S. value of 2.738 kWh /$ Nom.-PPP) [12].

Emissions in 1998 from (million metric tons): Solid fuels (172), liquid (25,410), gaseous (11,325), gas flaring (40,203) cement (1,345)

Carbon dioxide emission due to gas flaring: 3.3456kg of carbon dioxide per kWh per capita (relative to Nigerian population). The CO_2 emissions from energy sector are expected to grow by 2.2 % annually [13].

Thus, from the foregoing climatic characteristics and energy consumption mix in Nigeria, it is evident that the major generic causes of climate change are energy related activities.

3. Current status of Climate Change Convention in Nigeria

Nigeria submitted her first national communication under the Untied Nations Framework Convention on Climate Change (UNFCCC) in November 2003. The bold step to reduce hazardous emissions also set out / indicated the options/alternative for reducing emissions in the energy sector. These alternatives include:

a. Energy use efficiency improvement options in the industrial , residential and commercial sectors
b. Increased use of renewable resources, by introducing small scale hydropower plants and solar-electric options
c. Supply-side options, especially rehabilitation of some existing oil refineries and power plants, and the introduction of newer combined-cycle technologies and cogeneration at industrial and rural areas
d. Increased domestic use of associated natural gas to reduce gas flaring

Much of these lofty pathways to emission reduction are yet to be objectively implemented. Under the current status climate change problem in Nigeria could be worse that first thought. This implies that strong legislative enactment should be in place to regulate whatever adaptation polices being adopted.

3.1. Climate change risks of energy production and emissions inventory

Energy production and consumption is known to be associated with some hazards to the ecosystem. The hazardous effects results in environmental degradation. The consequences, which are more prominent in the destruction of surrounding vegetation and marine life, could cause enormous devastation of the surrounding environment generally. For example, in Nigeria, the impact of oil operations on the environment has produced a technological shock of unexpected dimensions [14].

The climate change risks associated with energy exploration and exploitation activities can be categorized according to stages involved in the energy activities. These stages are encountered during discovery, harnessing, processing, storage, transportation and final utilization of energy. With adequate knowledge of the risks through the aforementioned stages then the deleterious effects can be grouped as follows:

Harm to living organisms,
Hazard to human health,
Hindrance to marine lives,
Impairment to water and air qualities through pollution, and
Impairment to profitable agricultural production, etc.

Figure 5 shows pictorially climate risks associated with energy extraction and use as expressed in a climate change tree.

These risks are not one time events, but are likely to increase in frequency and intensity. Hence, adaptation mechanisms are needed for their resilience by beneficiaries.

However, effective emission inventory is critical to cost-effective adaptation strategy. The purpose of an emission inventory is to locate the air pollution sources for a given location and to define the types and magnitude of pollutants that these sources are likely to produce. It projects pollutants and their frequency, duration and relative contribution from each source. Thus, climate change adaptation and carbon dioxide emissions mitigation can significantly be achieved if pollutant inventory level is known with certainty.

4. Strategies for curbing climate change risks

In Nigeria, increase in the total energy production and consumption and the resulting impact of climate change will continue for a long time, because of the following:

- Energy resources keep the economy running
- 90 % of government revenue is derived from energy royalty
- Substitute for transport, heavy industrial and domestic energy needs with renewable is implausible in the distant future.

The strategies available to Nigeria are to develop trans-disciplinarity skills to adapt to climate change consequences. That is, a form of collaborative problem solving expertise involving science, technology and society (the ultimate beneficiary) in varying priorities as bulwark against climatic changes. The above necessitates the promotion of adaptation strategy as a trade-in strategy to balance climate change effects. Since the forcing agents of the change are expected to continue for reasons of more energy and human activities, domesticating the strategies through a combination of policy changes, technology innovations and best practices are necessary steps.

4.1. Policy, technology innovations, and best practices for climate change adaptation

The world cannot reverse climate change. The effects will continue for a long time. What ought to be done in the light of prevailing mindset on consequences of climate change is how to live with the changes. Thus, living with climate change implies that adaptation strategies are of utmost importance, in the short-run, though mitigation measures also are important for long term solutions and should be widely promoted. Let us examine, in this paper, innovative climate friendly policies and technologies for advancing of climate resilience and low-carbon development in Nigeria.

The climate friendly policies and technologies are broadly classified into educational policy and energy technology policy. On the energy policy strategy the highlights include:

- Cogeneration (cost-effective waste energy utilization)
- Energy planning and management (developing low ambient energy; promoting less dependence on fossil fuels, nanotechnology, photovoltaics, forestry).

The key points on focus for the educational strategies include:

- Capacity building (unifying the knowledge of the problem with sustainable solutions)
- Curriculum revision to incorporate learning of climate change (express in content the problem, learning objectives and feasible outcomes).

4.2. Energy technology policy

4.2.1. Waste energy utilization - Cogeneration

The key societal needs linked to energy utilization include:

- Residential and commercial buildings
- Air-conditioning and refrigeration
- Automotive propulsion
- Cement manufacturing
- Steel production
- Agriculture, fertilizer and processing
- Energy exploration and gas flaring
- Power generation

In these tasks, there are currently enormous energy wastes and opportunities for improvement. The wastes can be reduced, recycled and reused (3R) in further utility applications. For example, the on-going power sector reforms in Nigeria provide an opportunity to transform the sector in two ways: integration of cogeneration in the existing structure and scale-up renewable energy usage in the energy mix. What are the benefits and cost implications of the suggested transformation methods? These are common questions posed by sceptics of the aforementioned reform process.

Integrating cogeneration technology can achieve savings in cost monetarily and environmentally [15]. For now, we focus on the environmental savings. The choice of cogeneration technology is also influenced by its compatibility with available energy infrastructure in Nigeria [16].

The cogeneration concept describes the production of both electricity and thermal energy from same facility. In cogeneration, also called combined heat and power (CHP), the energy in the combusted fuel is used twice. The advantage of CHP results from capturing the waste heat created in the process of producing electric or thermal energy traditionally exhausted through the stack [17]. The secondary advantage of cogeneration is less fuel consumption that translates to less polluting of the environment. High level consumption of fossil fuels is responsible for greenhouse gas emissions that cause global warming. Reduced emission of greenhouse gases (nitrous oxide, carbon dioxide, methane, and water vapour) is possible with cogeneration technology. Taking full advantage of these potentials would lead to better economy of resources and friendly environment. Some characteristics of prime movers used for the cogeneration system as reported by Wu and Wang [18] are shown in table 1. However, application of cogeneration in Nigeria has ethical limitations.

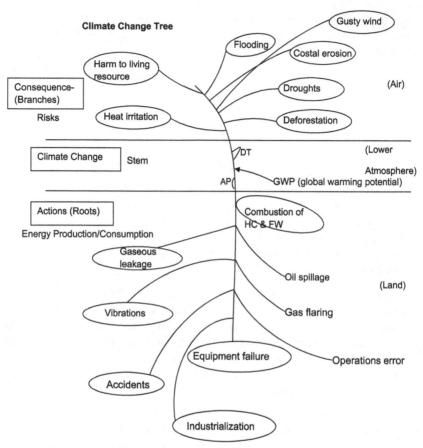

Keys: AP = acidification potential; DT = differential temperature; HC = hydrocarbon; FW= fuelwood.

Figure 5. Climate change tree illustrating the risks associated with energy production

In Nigeria, there is acute shortage of generated grid electricity. Under the circumstance, it is difficult to operate cogeneration effectively without a reliable supply end. Secondly, self-autogeneration is the second best option that is relied upon to supply energy to all facets of activity due to unreliable grid system. Operating cogeneration under this situation is counter productive. Self-autogeneration is known for its high emission of GHGs more than the grid system. Nevertheless, these are ethical and management issues (sabotage, lack of integrity, poverty related) and there are rooms for improvement [10].

4.2.2. Renewable energy and alternative fuels sources

These energy resources offer environmental benefits ranging from low carbon and sulphur emissions to non-emissions of greenhouse gases. The renewable solutions include wind

	Steam turbine	Spark ignition engines	Gas turbines	Micro-turbines	Stirling engines	Fuel cells
Capacity range	50kW–500MW	3kW – 6MW	250kW – 50MW	15kW – 300kW	1kW – 1.5MW	5kW – 2MW
Fuel used	Any	Gas, biogas, liquid fuels, propane	Gas, propane, distillate oils biogas	Gas, propane, distillate oils biogas	Any (gas, alcohol, butane, biogas)	Hydrogen and fuels containing hydrocarbons
Efficiency electrical (%)	7 – 40	25 – 43	25 – 42	15 – 30	~40	37 – 60
Efficiency overall (%)	60 – 80	70 – 92	65 – 87	60 – 85	65 – 85	85 – 90
Power to heat ratio	0.1– 0.5	0.5 – 0.7	0.2 – 0.8	1.2 – 1.7	1.2 – 1.7	0.8 – 1.1
CO_2 emissions (kg/MWh)	c	500 – 620	580 – 680	720	672[d]	430 – 490
NO_x emissions (kg/MWh)	c	0.2 – 1.0	0.3 – 0.5	0.1	0.23[d]	0.005 – 0.01
Availability (%)	90 – 95	95	96 – 98	98	N/A	90 – 95
Part load Performance	Poor	Good	Fair	Fair	Good	Good
Life cycle (year)	25 – 35	20	20	10	10	10 – 20
Average cost Investment (S/kW)	1000 – 2000	800 – 1600	450 – 950	900 – 1500	1300 – 2000	2500 – 3500
Operating and Maintenances cost ($/kWh)	0.004	0.0075– 0.015	0.0045 – 0.0105	0.01 – 0.02	N/A	0.007 – 0.05

Table 1. Characteristics and parameters of prime movers used in cogeneration system

energy, solar energy, geothermal systems, nuclear energy and biofuel / biomass. These sources have the capability to renew themselves, although the nuclear option is usually marred by public outcry as a result of the use of plutonium-239 in fast breeders. In biomass, fuelwood consumption is an aberration in standard energy mix due to poverty and deficient in conventional energy infrastructure. Fuelwood problem is solvable.

The alternative fuels such as biofuels are categorized as follows:

- First-generation biofuel (edible plants materials)
- Second-generation biofuels (non-edible plant materials)
- Third-generation biofuel (algae)

The use of biofuels does not contribute to global warming, as the CO_2 they release when burnt is equal to the amount that the plants absorb out of the atmosphere [19]. Why we cannot depend so much on renewable energies for climate change adaptation is because the technology is alien to most vulnerable developing countries. Their initial cost of acquisition is prohibitive.

4.2.3. Calculating greenhouse gas emission impacts

To calculate the value of CO_2 and other greenhouse gas emissions reduction achieved by renewable and alternative energy, the information on the amount of kWh or MWh produced is required for a given utility. Each energy utility is required to produce specific grams of greenhouse gases while delivering 1 kWh at given conditions of operation. To associate the calculated value with renewable system, we make assumption that it would emit an equivalent amount for 1 kWh generation.

A general guideline in the assumption is to use as much local data as is possible. The data maybe observed or derived from literature. The following things are important in greenhouse estimation:

- to calculate greenhouse gas emissions for assigned lifetimes of facility or project
- to convert greenhouse gas impacts to metric tons of carbon dioxide equivalent
- to obtain carbon dioxide equivalent intensity of fuel or energy under consideration
- to treat carbon dioxide equivalent reductions as cumulative reduction

The above mentioned guide will help to accurately estimate the amount of emission reductions generated from energy project.

4.3. Changes proposed in educational policies

4.3.1. Institutional curriculum review to fit in climate change knowledge

Revision of curricula of studies at all levels for inclusion of climate change adaptation is a necessity in developing countries. Curriculum can be defined as a document, plan or blue print for instructional guide used for teaching and learning to bring about positive and desirable learner behaviour change [20]. In this context, it is an institutional policy tool for bringing about or directing a desired change in a learning activity. The levels of the institution where the curricula changes are expected include primary, secondary and the tertiary with varying degree of climate science knowledge.

The process of integration involves the following approaches:

- Situational analysis (using questionnaire, interview techniques)
- Formulation of objectives (philosophy/rationale/motivation)
- Identification of resources(human, materials, intangible)
- Organization of the curriculum (sequence, scope)
- Evaluation of outcome(statistical, pictorial, reporting) [21]

Climate change integration into curricula of studies is critical to raising awareness and getting human reactions about the changing in weather conditions. This is a boost because lack of awareness is the main obstacle to vulnerability to climate change effects.

4.3.2. Capacity building in climate change

There is increasingly difficulty to get talents in developing countries that understand and can interpret the science of climate variability. Climate change science is a complex phenomenon. The ability to understand and effectively communicate potential future climate scenario to decision makers are thus critical success factors for adaptation strategies [21]. This capacity resides with human resources. It is impossible to adapt to climate change without linking the process up with people who take the decision that impairs or makes the environment.

Before getting ready to train the required capability, human resource audit should be conducted in the critical areas of need. The audit will look at a variety of human resource management functions: staffing, training, appraisal and development, and overall effectiveness [22]. Essentially, it will help to identify local content of manpower requirements for capacity building process and domestication of those competencies.

To train people for climate change adaptation is critical to the success of the campaign on environmental sustainability with climate change. Although the training should be goal-driven, the output is largely determined by the input of the trainee. Awareness and educational capacity reflect on the expected input of the trainee. Therefore, he should be exposed to challenges and allowed a robust level of latitude to contribute towards dealing with the problems. The robustness makes for creativity. However, the capacity building approach to climate change adaptation has a long gestation period as a setback.

Yet legislation alone cannot offer all the solutions of climate change adaptation. There is need to focus on the attitude and commitment of people who operate the policies. The future challenge for the educational and energy technology policies perhaps is implementation. The more the implementation is devoid of corrupt tendency the farther the project of advancing climate change resilience will be on its road to success.

To attain best practices in this policy changes:

- complete plan for data collection must be defined before the collection process is began
- integrity of the strategies should be guaranteed through transparency
- capacity building should not end with tackling climate change, but should open up windows of opportunities for job creation
- energy policy strategy should promote energy efficiency
- in-situ use and compression of gas are adequate for managing gas flaring
- massive tree planting programme

With these practices and full implementation of the suggested policies low-carbon development is guaranteed in Nigeria.

5. Conclusion

The study presents the climate change risks associated with energy production and utilization as well as the possibility of achieving low carbon development in Nigeria. It also describes that the climate change is a global phenomenon, but the adversity of its impacts depends on the types of causative agents and geographical locations of the beneficiary. Thus, following conclusions are drawn:

The major generic causes of climate change are energy related activities as energy consumption in Nigeria are found with mix ratio of 58 % oil, 34 % natural gas, 8% hydroelectricity, where in production 50% natural gas is flared.

The potent climate forcing agents are found from greenhouse gas emissions (such as: C_{02} , $C0$, S_{02} , CH_4 , and N_{0x}) and is from energy related resources.

The proposed changes in educational policies and application of cogeneration are found cost effective. If climate change adaptation strategies coupled with technological innovation, it will promote low-carbon development in Nigeria.

The best measures practices for realization of low carbon society are proposed to be adopted such as: transparency, capacity building and economic empowerment opportunities in energy sector.

The study also recommends that some ethical challenges are to be identified which affect the success in the framework of suggested policy e.g., sabotage and lack of integrity in policy implementation.

Author details

S.C. Nwanya
Department of Mechanical Engineering, University of Nigeria, Nsukka, Nigeria

6. References

[1] Contreras L. R., K. Cuba. The potential impact of climate change on the Energy sector in the Caribbean region, Sustainable Energy and Climate Change Division, Department of Sustainable Development Organization of American states; 2008.

[2] Arapatsakos, C., A. Karkanis, M. Moschou, I. Pantokratoras. The variation of gas emissions in an Otto engine by using different gases as fuel. International J. of Energy and Environment 2012; vol. 6 (1), 49 – 52.

[3] Nwanya, S. C. Climate Change and Energy Implications of Gas Flaring for Nigeria. International J. of Low-Carbon Technologies 2011; vol. 6 (3), 193 – 199.

[4] Urama, K. C. Concepts and Principles of Trans-disciplinarity, Systems Thinking and Innovation in African Universities. Presented at ATPS-OSF-TRECAFRICA TD Training Workshop, University of Nigeria Nsukka, 2012; 5th March.

[5] Intergovernmental Panel on Climate Change (IPCC). The report of working group 1 of the IPCC, Survey for Policymakers, 2001.

[6] Iloeje, O. C. Globalization, Cross- Border partnership and Networking. Proceedings of COREN 16th Engineering Assembly. 28- 29th August, Abuja, 2002; pp 81-91, Nigeria.

[7] U. Etim, U. Ituem, A. Folarin,. Niger Delta Region of Nigeria, Climate Change and the Way Forward. American Society of Agriculture and Biological Sciences 2008. Available from: http://asae.frymulti.com (accessed 6 August, 2010).

[8] Odjugo, P. A. O. Quantifying the cost of climate change impact in Nigeria: Emphasis on wind and rainstorm. J. hum. Ecol. 2009; 28 (2), 93 – 101.

[9] Nwaroh, J. Opportunities and challenges of gas flaring in the oil industry. Presented at The Nigerian Society of Engineers' International Conference on Engineering response in combating effects of climate change in Africa, December 2010; 6th – 10th, Abuja Nigeria.

[10] Kennedy-Darling, J., N. Hoyt, K. Murao, A. Ross. The Energy Crisis of Nigeria: an overview and implications for the future. Thesis submitted to The University of Chicago, USA 2008.

[11] Odjugo, P. A. O. Regional evidence of climate change in Nigeria. Journal of geography and Regional Planning 2010; 3(6) 142-150.

[12] Sambo, A. S. Industrial Energy Management Practices in Nigeria. Presented at the UNIDO EGM on Industrial Energy Efficiency and Energy Management Standards, 2007; 21- 22 March, Vienna.

[13] Federal Ministry of Environment (FME). Needs for Climate Change in Nigeria. Special Climate Change Unit, FME Abuja, Nigeria; 2010.

[14] Adinkwu, J. I. The environmental hazards associated with oil production: a case study of Warri, Nigeria. Unpublished M. ENG Thesis, University of Nigeria Nsukka, Nigeria; 2003.

[15] Nwanya, S. C. Analysis of Cogeneration powered Absorption Chiller Systems in Remote Tropical areas. Thesis (Ph.D). Universita degli Studi di Udine, Italy; 2008a. Available from: www.scrib.com/doc/36894195/Analysis (accessed 6 March, 2012).

[16] Nwanya, S. C., R. Taccani, M. Reini. Cogeneration and sustainable energy development in Nigeria. Proceedings of 1st National Engineering Conference on Sustainable Energy Development in Nigeria: challenges and prospects. University of Ado-Ekiti Nigeria, 7 - 9th October, 2008b.

[17] Beyene, A., Combined heat and power as a feature of Energy Park. Journal of Energy Engineering, ASCE, (2005)

[18] Wu, D. W., R. Z. Wang. Combined cooling, heating and power: a review. Progress in Energy and Combustion Science 2006; vol. 32, 459 – 495.

[19] Offorma, G. C. Integrating Climate Change in the Curriculum of the University of Nigeria. Presented at ATPS-OSF-TRECAFRICA TD Training Workshop, University of Nigeria Nsukka; 5th March, 2012.

[20] Blanton, B., J. McGee, O. Kapeljushnik. Communicating Coastal Risks Analysis in an Age of Climate Change. Renaissance Computing Institute, TR-11-04; October, 2011.

[21] Noe, R. A., J. R. Hollenbeck, B. Gerhart, P. M. Wright. Fundamentals of Human Resource Management. McGraw-Hill/ Irwin publishers, USA; 2007.

Chapter 6

Methods of Estimating Uncertainty of Climate Prediction and Climate Change Projection

Youmin Tang, Dake Chen,
Dejian Yang and Tao Lian

Additional information is available at the end of the chapter

1. Introduction

A critical issue in climate prediction and climate change projection is to estimate their uncertainty. The estimation of uncertainty has been an intensive research field in recent years, which is also called the potential predictability study. The terminology of the uncertainty of prediction and the potential predictability are often alternatively used in literature due to their inherent linkage, although they have some difference in a rigorous framework of predictability theory. For example, when a system has a high potential predictability, we may think the uncertainty of its predictions to be small, and vice versa. In this chapter, unless otherwise indicated, the uncertainty of prediction and potential predictability have the similar meaning in describing and measuring the prediction utility, and are thus used alternatively. For simplicity, we also often use the term of predictability to denote the potential predictability.

The uncertainty of prediction or predictability study is usually conducted using the strategy of ensemble prediction, from which there are a couple of metrics to quantify the potential predictability. Among them are variance-based measure and information-based measure, both quantifying the predictability or prediction uncertainty from different perspectives. In this chapter, we will introduce the two kinds of metrics. Emphasis will be placed on the similarity and disparity of these measures, and the realistic applications of the measures in studying the uncertainty of climate prediction and climate change projection. It should be noted that these potential predictability metrics do not make use of observation, which is essentially different from the actual prediction skills measured against observations like correlation skill or root mean square of errors (RMSE).

2. Two methods of measuring potential predictability

2.1. Signal-to-Noise Ratio (SNR) and potential predictability

The SNR has been a widely used measure of potential predictability [1, 2]. At seasonal time scale, the signal is usually regarded as the atmospheric responses to the slowly varying external forcing such as sea surface temperature (SST), sea ice, snow cover, etc., whereas the noise is induced by the relatively high frequency atmospheric variability such as weather processes. In an ensemble seasonal climate prediction, the amplitude of signal and noise can be approximately quantified by the variance of ensemble mean and the averaged ensemble spread over all initial conditions [3-5], namely,

$$Var(S) = \frac{1}{M}\sum_{i=1}^{M}\left(\overline{X}_i - \overline{\overline{X}}\right)^2 \tag{1}$$

$$Var(N) = \frac{1}{MK}\sum_{i=1}^{M}\sum_{j=1}^{K}\left(X_{i,j} - \overline{X}_i\right)^2 \tag{2}$$

where $X_{i,j}$ is the j-th member of the ensemble prediction starting from the i-th initial condition. The X itself can be a scalar such as an index or a two dimensional field. K is the ensemble size and M is the total number of initial conditions (predictions); and $\overline{X}_i = \frac{1}{K}\sum_{j=1}^{K}X_{i,j}$, $\overline{\overline{X}} = \frac{1}{M}\sum_{i=1}^{M}\overline{X}_i$.

Considering the sampling errors in estimating signal variance, the more accurate estimation of signal variance S is modified as below:

$$Var(S) = \frac{1}{M}\sum_{i=1}^{M}\left(\overline{X}_i - \overline{\overline{X}}\right)^2 - \frac{1}{K}Var(N)$$

Two common measures of potentially predictability are the signal-to-noise ratio (SNR) and the signal-to-total variance ratio (STR), i.e.,

$$SNR = \frac{Var(S)}{Var(N)}, \quad STR = \frac{Var(S)}{Var(S) + Var(N)} \tag{3}$$

It can be derived that the square root of STR is equivalent to the correlation of the signal component (S) to the prediction target itself. Thus, the $\sqrt{STR}$ is often defined as potential correlation (PCORR).

It is easy to derive that the $\sqrt{STR}$ is actually a perfect correlation skill, which assumes that the observation is an arbitrary ensemble member. The perfect correlation skill ignores the imperfectness of model itself. To see this equality, we denote the ensemble mean μ as the prediction, thus the 'observation' can be written by $\mu + \varepsilon$, where the ε is a normally distributed white noise with the mean of zero and variance of σ_e^2.

The correlation of prediction against the 'observation' can be written as follows:

$$Corr_{pef} = \frac{E[\mu(\mu+\varepsilon)]}{\{E(\mu^2)E[(\mu+\varepsilon)^2]\}^{1/2}} = \frac{E(\mu^2)}{\sqrt{E(\mu^2)E[\mu^2+2\mu\varepsilon+\varepsilon^2]}} = \frac{\sqrt{E(\mu^2)}}{\sqrt{E(\mu^2)+E(\varepsilon^2)}} \tag{4}$$

Comparison between (3) and (4) reveals the equality of $Corr_{pef}$ and $\sqrt{STR}$.

2.2. Information-based potential predictability

2.2.1. Relative Entropy and predictive information

Entropy is a measure of dispersion level (e.g. uncertainty). The entropy of a continuous distribution $p(x)$ is defined as

$$H(x) = -\int p(x)\ln p(x)dx,$$

where the integral is understood to be a multiple integral over the domain of x. Larger entropy is associated with smaller probability and larger uncertainty.

The information-based potential predictability measures include relative entropy (RE), predictive information (PI) and predictive power (PP). The central idea of these information-based measures is that the difference between two probability distributions: the forecast distribution and climatology distribution, quantifies the extra information brought from the prediction.

Suppose that the future state of a climate variable is predicted/modeled as a random variable denoted by v with a climatological distribution $p(v)$. One ensemble prediction produces a forecast distribution which is the conditional distribution $p(v|i)$ given the initial condition i . The climatological distribution is also the unconditional distribution and we have

$$p(v) = \int p(v|i)p(i)di, \tag{5}$$

where the $p(i)$ is the probability distribution of the initial condition i . Usually various statistical tests are used to examine the difference between two distributions [6-7]. Relative entropy RE, or Kullback-Leibler distance, is a quantitative measure of the difference between two distributions from information theory [8]. In the context of predictability, it is defined as

$$RE = \int p(v|i)\ln\frac{p(v|i)}{p(v)}dv . \tag{6}$$

In terms of information theory, the quantity RE measures the informational inefficiency of using the climatological distribution $p(v)$ rather than the forecast distribution $p(v|i)$ and $RE \geq 0$ with the equality if and only if $p(v|i) = p(v)$ [8]. In Bayesian terminology, the climatological distribution is *a prior* distribution which can be usually derived from the long term historical observations. An ensemble prediction augments this prior information, and the additional information measured by RE is a natural measure of the utility or usefulness of this prediction and thus implies the potential predictability. In practice, $p(v|i)$ and $p(v)$ can be estimated directly from samples or approximated alternatively using kernel density estimation.

Another natural measure of predictability is the predictive information (PI), defined as the difference between the entropy of the climatological and forecast distributions:

$$PI = H(v) - H(v|i) \tag{7}$$

Considering (7), then

$$PI = -\int p(v)\ln[p(v)]dv + \int p(v|i)\ln[p(v|i)]dv \tag{8}$$

The first term on the right hand side of Eq. (8) denotes the entropy of the prior distribution $p(v)$ (climatological distribution), measuring the uncertainty of a prior time when no extra information is provided from the observed initial condition and forecast model; whereas the second term represents the entropy of the posterior distribution $p(v|i)$ (forecast distribution), measuring the uncertainty after the observed initial condition and subsequent prediction becomes available (An elaborated illustration can be found in [9]). Thus a large PI indicates that the posterior uncertainty will decrease because of useful information being provided by a prediction (e.g., the larger $p(v|i)$ the smaller uncertainty) that is, the prediction is to be more reliable in a "perfect model" context.

The predictive power (PP) was defined by [10]

$$PP = 1 - \exp(-PI) \tag{9}$$

In the case where the PDFs are Gaussian distributions, which is a good approximation in many practical cases (including ENSO prediction, e.g., [11]). The predictive and climatological variances, and the difference between their means. The resulting analytical expression for the relative entropy *RE, PI and PI* are given as follows [1]:

$$RE = \frac{1}{2}\left\{ \ln\left[\frac{\det(\Sigma_q^2)}{\det(\Sigma_p^2)}\right] + trace\left[\Sigma_p^2(\Sigma_q^2)^{-1}\right] + (\mu_p - \mu_q)^T(\Sigma_q^2)^{-1}(\mu_p - \mu_q) - n \right\} \tag{10}$$

$$PI = \frac{1}{2}\ln\left(\frac{\det(\Sigma_q^2)}{\det(\Sigma_p^2)}\right) \tag{11}$$

$$PP = 1 - \left(\frac{\det(\Sigma_q^2)}{\det(\Sigma_p^2)}\right)^{-\frac{1}{2}} \tag{12}$$

where, q and P are the climatological and predictive covariance matrices respectively; det is the determinant operator and tr is the trace operator; μ_q and μ_p are the climatological and predictive mean state vectors of the system, and n is the number of degree of freedom;. RE is composed of two components: (i) a reduction in climatological uncertainty by the prediction [the first two terms plus the last term on the right-hand side of (10)] and (ii) a difference in the predictive and climatological means [the third term on the rhs of (10)]. These components can be interpreted respectively as the dispersion and signal components of the utility of a prediction[12]. A large value of RE indicates that more information that is different from the climatological distribution is being supplied by the prediction, which could be interpreted as making it more reliable [1]. A key difference between relative entropy (RE) and predictive information (PI) is that RE vanishes if and only if the forecast and climatological distributions are identical (i.e., same mean and spread), while PI is zero as long as the two distributions have the same spread [9]. Remarkably, predictive information and relative entropy are invariant with respect to linear invertible transformations of the state [9-10].

For a scalar variable (e.g., an index), RE, PI, and PP can be simplified as

$$PI = \frac{1}{2}\ln\left(\frac{\sigma_q^2}{\sigma_p^2}\right) \tag{13}$$

$$RE = \frac{1}{2}\left[\ln\left(\frac{\sigma_q^2}{\sigma_p^2}\right) + \underbrace{\frac{\sigma_p^2}{\sigma_q^2} - 1}_{Dispersion} + \underbrace{\frac{\mu_p^2}{\sigma_q^2}}_{Signal}\right] = PI + \frac{1}{2}\left[\frac{\sigma_p^2}{\sigma_q^2} - 1 + \frac{\mu_p^2}{\sigma_q^2}\right] \tag{14}$$

$$PP = 1 - \left(\frac{\sigma_p^2}{\sigma_q^2}\right)^{1/2} \tag{15}$$

2.2.2. Mutual information

RE or PI is a predictability measure for individual predictions. The average of REs or PIs over all initial conditions reflects the average predictability and was proved to be equal to mutual information (MI), another quantity from information theory [9]. In the context of predictability, MI is defined as [9]

$$MI = \iint p(v,i)\ln\left[\frac{p(v,i)}{p(v)p(i)}\right]dvdi \tag{16}$$

where $p(v,i)$ is the joint probability distribution between v and i. MI measures the statistical dependence between v and i, and vanishes when v and i are independent ($p(v,i) = p(v)p(i)$).The equality of MI and average RE indicates that predictability can be measured in two equivalent ways: by the difference between forecast and climatological distributions or by the degree of statistical dependence between the initial condition i and the future state v [13]. If the future state v is on average unpredictable, individual forecasts should have the probability distribution identical to the climatological distribution, i.e., $p(v \mid i) = p(v)$ and RE $=0$ for all predictions. This is equivalent to independence between i and v . Therefore, independence indicates unpredictability and dependence implies predictability. MI is invariant with respect to nonlinear, invertible (nonsingular) transformations of state[9]. Thus, the MI between v and i equals to the MI between v and ensemble mean $\mu_{v|i}$. The latter is probably more straightforward in understanding MI-based predictability since the dependence between v and ensemble mean $\mu_{v|i}$ can be interpreted as the dependence between observation (v) and prediction ($\mu_{v|i}$) under the assumption of a perfect model.

When forecast and climatological distributions are Gaussian, MI can be expressed, using (13), by[13]

$$MI = \frac{1}{2}\left(\ln \sigma_v^2 - \left\langle \ln \sigma_{v|i}^2 \right\rangle\right) \tag{17}$$

Eq. (17) is the formula often used to calculate MI. *Joe* [14] and *DelSole* [15] showed that the transformations $\sqrt{1 - \exp(-2MI)}$ and $1 - \exp(-2MI)$ produce "potential" skill scores which exhibit proper limiting behavior: they have values between 0 and 1, and the minimum (maximum) value 0 (1) occurs when MI vanishes (approaches infinite). Here "potential" indicates that they are perfect model measures. In this study, we will use the two "potential" skills to represent MI. Furthermore, if the forecast and climatological distributions are Gaussian, and forecast variance is constant, the above two "potential" skills respectively reduce to another two conventional "potential" skills: "potential" anomaly correlation (AC_p) and "potential" mean square skill score ($MSSS_p$)[9,13].

$$AC_p = \sqrt{1 - \exp(-2MI)}, \tag{18}$$

$$MSSS_p = AC_p{}^2 = 1 - \exp(-2MI) \tag{19}$$

2.3. Relationship between SNR-based metrics and MI-based metrics

The averaged RE and PI ($\overline{RE}$ and $\overline{PI}$) over all predictions (initial conditions) are identical to MI, as mentioned before. For seasonal climate prediction, the total variance (i.e., climate variance) can be decomposed into signal (S) variance and noise (N) variance, if the signal and noise are assumed to be independent of each other ([16-17]), namely,

$$Var(T) = Var(S) + Var(N) \tag{20}$$

where

$$Var(T) = \frac{1}{MK}\sum_{i=1}^{M}\sum_{j=1}^{K}\left(X_{i,j} - \overline{\overline{X}}\right)^2,$$

$$var(S) = \frac{1}{M}\sum_{i=1}^{M}(\overline{X}_i - \overline{\overline{X}})^2$$

$$Var(N) = \frac{1}{MK}\sum_{i=1}^{M}\sum_{j=1}^{K}\left(X_{i,j} - \overline{X}_i\right)^2.$$

$X_{i,j}$ is the j-th member of the ensemble prediction starting from the i-th initial condition. The K is the ensemble size and M is the total number of initial conditions (predictions); and

$$\overline{X}_i = \frac{1}{K}\sum_{j=1}^{K}X_{i,j}, \quad \overline{\overline{X}} = \frac{1}{M}\sum_{i=1}^{M}\overline{X}_i.$$

Without the loss of generality, the climatological mean is assumed to be zero, thus (20) can be expressed by

$$\overline{\sigma_q^2} = \overline{\mu_p^2} + \overline{\sigma_p^2} \tag{21}$$

where the overbar denotes the expectation over all predictions (initial conditions).Eq (14) and Eq. (21) can easily verify the property of MI, for example,

$$MI = \overline{RE} = \overline{PI} + \frac{1}{2}\left[\frac{\overline{\mu_p^2} + \overline{\sigma_p^2} - \overline{\sigma_q^2}}{\overline{\sigma_q^2}}\right] = \overline{PI} \tag{22}$$

Using (21), the information-based potential predictability measures MI, ($\overline{RE}$ or $\overline{PI}$) and $\overline{PP}$ can be rewritten as the function of the mean signal and noise, or their ratio SNR. The σ_q and σ_p in (21) are actually σ_v and $\sigma_{v|i}$ in (17), thus we have [18]

$$MI = \frac{1}{2}\left(\ln\sigma_v^2 - \left\langle\ln\sigma_{v|i}^2\right\rangle\right) \geq \frac{1}{2}\left(\ln\sigma_v^2 - \ln\left\langle\sigma_{v|i}^2\right\rangle\right)$$
$$= -\frac{1}{2}\ln\left(\frac{\left\langle\sigma_{v|i}^2\right\rangle}{\sigma_v^2}\right) = -\frac{1}{2}\ln(\frac{\overline{\sigma_p^2}}{\overline{\sigma_q^2}}) = -\frac{1}{2}\ln(1 - STR) \tag{23}$$

The inequality in (23) is due to the fact that arithmetic mean is larger than or equal to geometric mean, or more strictly is a result of Jensen's inequality from information theory. Therefore, we have

$$AC_p \leq \sqrt{1 - \exp(-2MI)}, \tag{24}$$

$$MSSS_p \leq 1 - \exp(-2MI).$$ (25)

The equalities in (24), (25) and (23) hold if and only if $\sigma_{v|i}^2$ is constant, as addressed in (18) and (19). The conditions that the forecast and climatological distribution are both Gaussian and the forecast variance $\sigma_{v|i}^2$ is constant are equivalent to the condition that i and v are joint normally distributed [13,18]. If i and v are joint normally distributed, the probability distributions $p(i)$, $p(v)$ and $p(v|i)$ are all Gaussian distributions and there are [13,18-19]

$$\mu_{v|i} = \rho_0 \frac{\sigma_v}{\sigma_i}\left(i - \langle i \rangle\right),$$ (26)

$$\sigma_{v|i}^2 = (1 - \rho_0^2)\sigma_v^2 = constant,$$ (27)

$$MI = -\frac{1}{2}\ln(1 - \rho_0^2),$$ (28)

where(26) is obtained using a linear regression with ρ_0 being the linear correlation between the initial state i and the future state v. As mentioned earlier, MI measures the statistical dependence between i and v. As can be seen from (26), the statistical dependence reduces to a linear correlation ρ_0 if the two variables are joint normally distributed. Because conditional mean $\mu_{v|i}$ is a linear function of initial state i (see (26)), ρ_0 is also the linear correlation between $\mu_{v|i}$ and v, which is the potential anomaly correlation skill AC_p. Note that if i and v are not joint normally distributed, ρ_0 is usually different from AC_p.

One interesting question arises here, namely that, how we understand the MI-SNR discrepancy when there is significant variability of prediction variance, as expressed in (24) and (25)? As discussed earlier, the MI-based potential predictability measures the statistical dependence, liner or nonlinear, between the ensemble mean prediction $\mu_{v|i}$ and the hypothetical observation v (an arbitrary ensemble member), whereas the SNR-based potential skill only measures their linear correlation. When $\mu_{v|i}$ and v are joint normally distributed, their statistical dependence reduces to linear correlation. When $\mu_{v|i}$ and v are not joint normally distributed, MI naturally disagrees with SNR. The joint normally distributed variables have constant conditional variance. Note that prediction variance is also the conditional variance of v given the ensemble mean $\mu_{v|i}$. Thus, if the prediction variance is varied, $\mu_{v|i}$ and v are definitely not joint normally distributed, making the SNR-based potential skill, a linear correlation between $\mu_{v|i}$ and v, underestimate the nonlinear statistical dependence between $\mu_{v|i}$ (or i) and v, which is a strict statistical definition of potential predictability.

It should be noted that the above conclusion should not be challenged by a possible fact that SNR-based skill might have a better relationship to actual skill than MI-based skill, simply because the actual skill is often measured by the linear correlation (or related quantity), which is inherent to the SNR-based skill. Thus, a more challenging issue is how to design new metrics to measure actual forecast skill which could appreciate the MI-based extra predictive information beyond SNR. In principle, the MI between ensemble mean prediction μ_{vli} and actual observation O could have the potential capability to quantify the MI-based potential predictability [15]. However, how to effectively estimate MI in this context is not an easy issue.

In summary, there are connections between information-based potential predictability and SNR-based potential predictability, as built by the above equations. In other words, all the averaged information-based potential predictability measures are better than SNR-based predictability in characterizing 'true' potential predictability. When the climatology and prediction distribution are both Gaussian and the prediction variances are constant, the information-based measure is equivalent to the SNR-based potential measure.

3. Maximum SNR and PrCA

The signal and noise are theoretically statistically irrelevant when the ensemble size is infinite. However, the ensemble size is always finite in reality, thus the estimation of the signal is often contaminated by the noise. An optimal estimate for the largest potential predictability should be to maximize the SNR, from which the resultant signal component is the most predictable.

We denote by $\mathbf{S}$ and $\mathbf{N}$ signal and noise of variable $\mathbf{X}$, where $\mathbf{S}$ and $\mathbf{N}$ are matrixes of a two-dimension field describing temporal and spatial variation of the signal and noise of one variable of interest, namely, this section is at the framework of the multivariate statistics.

where, $\mathbf{S} = \overline{X}_i - \overline{\overline{X}}$; $\mathbf{N} = X_{i,j} - \overline{X}_i$

$X_{i,j}$ is the j-th member of the ensemble prediction starting from the i-th initial condition. The K is the ensemble size and M is the total number of initial conditions (predictions); and

$$\overline{X}_i = \frac{1}{K}\sum_{j=1}^{K} X_{i,j}, \quad \overline{\overline{X}} = \frac{1}{M}\sum_{i=1}^{M} \overline{X}_i.$$

Our goal is to look for a vector q, which can maximize the ratio of the variance of signal and noise that are projected onto the vector, namely,

$$r_S = q^T * \mathbf{S}; \quad r_N = q^T * \mathbf{N}$$

$$\text{SNR} = \frac{\sigma^2_{r_S}}{\sigma^2_{r_N}} \quad ==> \quad \text{max} \tag{29}$$

where

$$\sigma_{r_s}^2 = \frac{1}{M} r_s * r_s^T = \frac{1}{M} q^T S * S^T q = q^T \Sigma_S q$$

$$\sigma_{r_N}^2 = \frac{1}{KM} r_N * r_N^T = \frac{1}{KM} q^T N * N^T q = q^T \Sigma_N q$$

Mathematically such an optimization by (29) leads to a generalized eigenvalue-eigenvector problem based on the Rayleigh Quotient theorem[10,20]

$$q^T \Sigma_S = \lambda q^T \Sigma_N \qquad (30)$$

Where, Σ_S is the covariance matrix of signal, Σ_N is the covariance matrix of noise. The solution of (30) can be obtained by solving the below eigenvalue equation

$$q^T \Sigma_S \Sigma_N^{-1} = \lambda q^T \qquad (31)$$

Thus, the analysis of the largest potential predictability is also called the maximum signal-to-noise EOF (MSN EOF) analysis, first introduced by Allen and Smith [21] to estimate the signal optimally by suppressing the influences of noise, and widely used already in climate predictability study [20,22-23].

Practically, the number of grid points is always much larger than the number of total samples in climate studies, thus usually Σ_N doesn't have full-rank, leading to a solution of ill-conditioned inversions. There are two common methods to solve this issue, as introduced below.

3.1. The SNR is optimized in a truncated EOF space

Denote by e_i^T (i=1,2...,k) the EOF modes[1] , and the signal and noise components projected on them are

$$T_S = e_i^T * S$$

$$T_N = e_i^T * N \qquad (32)$$

T_S and T_N are PC components with dimension of k*n where the n is the time samples. Thus, the signal and noise variance used in (31) should be calculated by $\Sigma_S = T_S * T_S^T$, $\Sigma_N = T_N * T_N^T$, respectively. If k truncated modes remain, where k is usually much smaller than the number of spatial grids, the signal and noise covariance matrix is a full-rank of k matrix. Thus, eq. (31) can be easily solved and the vector q (denoted as q_{eof}) is called filter pattern, which is a k-element vector, the filter pattern on the truncated EOF space. The leading predictable component is

---

[1] e_i is a matrix of m*k, where m is the number of spatial grids and k is the number of the truncated modes. EOF could be employed using signal, or noise matrix or corresponding data matrix.

$$r_S = q_{eof}^T * T_S \tag{33}$$

If projecting q_{eof} back to data space, we have

$$PrC\_data = q^T * S = q_{eof}^T * e^T * S = q_{eof}^T * T_S \tag{34}$$

Eq. (33) and (34) are identical each other, i.e., PrC is invariant with respect to a linear transformation.

The q is the filter pattern, rather than the most predictable pattern. The most predictable pattern v can be obtained using the regression method, i.e.,

$$N = r_N * V \tag{35}$$

$$V = N * r^T{}_N / (r_N * r_N^T) = \frac{1}{MK} N * r^T{}_N$$

$$= \frac{1}{MK} N * T_N^T * q$$

$$= \frac{1}{MK} N * N^T * e * q$$

$$= \Sigma_N * e * q$$

Also it can be written by

$$S = r_S * V \tag{36}$$

$$V = S * r_S^T / r_S * r_S^T =$$

$$\frac{1}{M\lambda} S * r_S^T =$$

$$\frac{1}{M\lambda} S * S^T * e * q = \frac{1}{\lambda} \Sigma_S * e * q$$

$$= \Sigma_N * e * q$$

3.2. Solving eq (29) using whitening approach

The approach is to whiten the noise variance (i.e., the denominator), making Σ_N an identity matrix and whitening the covariance matrix of signal(Σ_S) simultaneously. Thus, eq. (29) becomes

$$SNR = \frac{q^T \Sigma_S q}{q^T \Sigma_N q} = \frac{q'^T \Sigma_{WS} q'}{q'^T q'} \rightarrow maximum \tag{37}$$

Based on the matrix theory, the SNR in eq. (37) reaches maximum when q' is the eigenvector of Σ_{WS}, the whitened signal covariance associated with the whitening noise Σ_N. The q' is a modified q by a whitening factor. The algorithm is briefly summarized as follows:

i.   Make the covariance matrix of noise (Σ_N) identity, namely,

$$D^{-1/2}E^T\Sigma_N ED^{-1/2} = I \tag{38}$$

D and E are the eigenvalue and eigenvector matrices of Σ_N. $ED^{-1/2}$ is the transformation matrix that makes the covariance matrix of noise (Σ_N) identity.

ii. Whiten the signal covariance matrix by the transformation matrix $ED^{-1/2}$, using the k leading modes

$$\Sigma_{WS} = D^{-1/2}E^T\Sigma_S ED^{-1/2} \tag{39}$$

iii. The SNR of (37) reaches maximum when q' is the leading eigenvector of the whitened signal covariance matrix Σ_{WS} (in descent order). It is easy to see the relationship between q and q'

$$q'^T\Sigma_{WS}q' = q'^T D^{-1/2}E^T\Sigma_S ED^{-1/2}q' = q^T\Sigma_S q \tag{40}$$

Thus,

$$q = ED^{-1/2}q'$$

iv. After the filter pattern q is known, the most predictable component is easy to derive as shown in method 1, namely projecting the signal (ensemble mean) on the filter patterns

$$\begin{aligned} PrCs\_s &= q^T S \\ PrCs\_n &= q^T N \\ PrCs\_t &= q^T T \end{aligned} \tag{41}$$

The most predictable component is the one corresponding to the largest signal-to-noise ratio. All PrCs are temporally orthogonal (uncorrelated) with each other. It is noted that (41) is a little different from (33) or (34) where the PCs of truncated EOF spaces are used. It is due to a different truncation procedure in the two methods. In this first method, the truncation is applied before optimization whereas in the second method, the truncation is implicitly integrated into the whitening process. However both should be equivalent, which can be seen by another expression of (41)

$$PrCs\_s = q^T S = q'^T D^{-1/2}E^T S = q'^T D^{-1/2}T_S$$

v.   Obtain the corresponding predictable patterns V by,

$$V = N * PrCs\_n^T/PrCs\_n*PrCs\_n^T = \frac{1}{MK}N*PrCs\_n = \frac{1}{MK}N*N_N^{T}*q = \Sigma_N q$$

A reconstructed forecast based on PrCS leading modes can be obtained by

$$
\begin{aligned}
\hat{N} &= V * \mathrm{PrCs\_}n \\
\hat{S} &= V * \mathrm{PrCs\_}s \\
\hat{X} &= V * \mathrm{PrCs\_}T
\end{aligned}
\tag{42}
$$

$\hat{X}$ only remains the leading PrCS modes and removes noise components, thus it can be expected to have a better skill than simple ensemble mean.

The variance explained by a PrCA mode can be obtained using (42). If all modes are remained in (42), the reconstructed filed should explain 100% of original field. We rewrite (42), applied into signal and noise, thus,

$$
\Sigma_S = \frac{1}{M} S^T * S; \quad \hat{\Sigma}_S = \frac{1}{M} V * r_S^T * r_S * V^T = V \Lambda V^T = \sum_{j=1} v_j * \lambda_j * v_j^T
$$

$$
\Sigma_N = \frac{1}{MK} N^T * N; \quad \hat{\Sigma}_N = \frac{1}{MK} V * r_N^T * r_N * V^T = V * V^T = \sum_j v_j * v_j^T
$$

Where the $\hat{\Sigma}$ is the estimated variance using PrCA modes. Thus, the variance explained by a specific mode measured in the original space, and the truncated space, is respectively as below:

$$
\text{relative to signal}: \quad \frac{\lambda_j v_j * v_j^T}{tr(\Sigma_S)} \quad ; \quad \frac{\lambda_j v_j * v_j^T}{tr(\hat{\Sigma}_S)}
$$

$$
\text{relative to noise}: \quad \frac{v_j * v_j^T}{tr(\Sigma_N)} \quad ; \quad \frac{v_j * v_j^T}{tr(\hat{\Sigma}_N)}
$$

$$
\text{relative to total variance}: \quad \frac{\lambda_j v_j * v_j^T + v_j * v_j^T}{tr(\Sigma_S) + tr(\Sigma_N)} \quad ; \quad \frac{\lambda_j v_j * v_j^T + v_j * v_j^T}{tr(\hat{\Sigma}_S) + tr(\hat{\Sigma}_N)}
$$

4. Maximizing PI and PrCA

Another interpretation to MSN EOF is its connection with information-based measure PI or PP defined in (11) - (15). For example, as argued in [10], the predictive power PP is a positively oriented predictive index, defined by the difference between posterior (prediction) entropy and prior (climatology) entropy, thus measuring the decrease of uncertainties due to prediction.

The PrCA analysis is an approach to maximize PI, or maximum PP, equivalent to minimizing σ_p^2 / σ_q^2 if the prediction variance is little changed, to derive the most predictable component. The σ_q^2 is climatology variance, often referred as to the total variance Var (T), which is composed of the signal variance and noise variance. Under the 'perfect model' assumption, the noise variance equals to the forecast error variance [24], namely,

$$\sigma_q^2 = Var(S) + Var(N) = Var(S) + \sigma_p^2 \tag{43}$$

Thus, the minimization of σ_p^2 / σ_q^2 is equal to the maximization of $1 - \sigma_p^2 / \sigma_q^2$, i.e., STR, which is equivalent to the maximization of SNR, i.e., MSN EOF. In some literature, the term of MSN EOF and PrCA are alternatively used due to their complete equivalence. Actually, both the MSN EOF and PrCA methods belong to the discriminant analyses because the two methods, though from different perspectives, can be understood to seek a best linear combination of variables that separates the signal and the noise as much as possible [13]. The both methods identify the "filter pattern", or weight matrix, providing an optimized filter to discriminate the signal and noise, where the time series reflects the temporal evolution of the dominant mode of the signal, and the spatial pattern characterizes the spatial distribution of the dominant mode of signal, which are respectively referred to as spatial pattern, or the most predictable pattern.

It should be noted that the equivalence of SNR-based and information-based PrCA approach is based on the condition that the climatology and forecast distribution are both Gaussian. It is apparent since the PI and PP cannot be only expressed by the form of prediction and climatology variance as (11) – (15) under non-Gaussian assumption. It is difficult to derive the optimization solution for PI or PP from their general definitions of (8) and (9).

A remark to the algorithm of Schneider and Griffies [10] is a technical issue. In Schneider and Griffies [10], the PrCA is proposed to derive by minimizing PP, i.e., minimizing σ_p^2 / σ_q^2, leading to the below eigenvalue equation:

$$f^T \Sigma_N \Sigma_T^{-1} = \xi f^T \tag{44}$$

where Σ_T is the total variance. The optimal filter resulting in the most predictable is the eigenvector f with the smallest eigenvalue ξ. Comparing (44) with (31) reveals that eigenvalue ξ and q are reciprocal, indicating the equivalence of PrCA using maximization of (31) and minimization of (44). Usually, the eigenvector with the smallest eigenvalue often lacks of a stable, large scale-like pattern, making the approach of (44) impractical in real application. The truncated EOF space, which is used in solving (31) and (44), can greatly reduce this concern but still the most predictable pattern contains some noise. Thus, the MSN EOF approach, introduced above, is a better option.

5. A practical application – Potential predictability of climate change projection in AR5

In this section, we will explore the uncertainty of climate change projection using the above theoretical framework. The estimation of uncertainty is based on the Coupled Model Intercomparison Project Phase 5 (CMIP5), a new set of climate model experiments involved in the Intergovernmental Panel on Climate Change (IPCC) Fifth Assessment Report (AR5). The CMIP5 is promoted to address some crucial issues on climate modeling and future climate state. More than 20 climate models were employed in this project with main focus on: 1) evaluate model predictability of future climate on different time scales (near term (out to about 2035) and long term (out to 2100 and beyond)), 2) understanding key mechanisms responsible for differences in model projections, and 3) quantify some important feedbacks of climate system like clouds and carbon cycle.

One of experiments used in CMIP5 is the Representative Concentration Pathways (RCPs) scenario. All model experiments involved in this scenario are forced by four kinds of mixing greenhouse gases (GHGs) boundary conditions which will finally lead increasing of radiation by 2.6, 4.5, 6.0 and 8.0 watt per square at the end of 21 century.

In this chapter, we will use the sea surface temperature (SST) projection of scenario R60 (the increasing of 6.0 watt per square experiment) to evaluate the potential predictability of climate projection of the scenario R60. At present, only nine models collected in R60 are available to download (From ESG-PCMDI Gataway), as summarized in table 1.

Model	Country	Ocean Model Resolutions	Projection
CCSM4-version16	USA (NCAR)	60 levels; 1.0 lon. x 0.5 lat.	2051-2100
CSIRO-MK3.6.0	Australia (BMRC)	30 levels; 1.875 lon x 0.9375lat.	2051-2100
GISS-E2-R	USA (NASA)	32 levels;1 lon x 1.25 lat.	2051-2100
GFDL-ESM2M	USA (GFDL)	50 levels; 1- 1/3 lon. x 1 lat.	2051-2100
HadGEM2-ES	UK (Hadley Center)	40 levels;1-1/3 lon. x 1 lat.	2051-2100
CM5A-LR	France (IPSL)	ORACA2 resolution in OPA	2051-2100
MIROC5-Coco 4.5	Japan	Varied resolution	2051-2100
MRI-CGCM3	Japan	Varied resolution	2051-2100
NorESM1-M	Norway	Varied resolution	2051-2100

Table 1. Models used for evaluation

The SST outputs from these models are all monthly averaged data. For the purpose of the study of the climate change, we use annual mean in the following discussions. Because the lack of uniformity of ensemble member, only one member is used for each model here. In this study, we confine the domain to the Pacific across 60S to 60N.

Shown in Fig. 1 and Fig 2a are the spatial pattern and time series of the first EOF (Empirical Orthogonal Function) for the Pacific Ocean from 2051-2100. As can be seen in Fig.2a, the Pacific SST has a striking increase, with the strongest response to the forcing of GHGs in the tropical Pacific along the equator as shown in Fig.1. In the extra-tropical beyond the 30S and 30N, the increase in SST is relatively weaker. On average, the mean temperature of the Pacific ocean of 60S to 60N increases around 0.5 to 1C from 2051-2010 in these models, as shown in Fig. 2b, the evolution of the mean temperature over the Pacific ocean. The mean of multiple models has the increase rate of around 0.75C as shown by the red line in Fig. 2b.

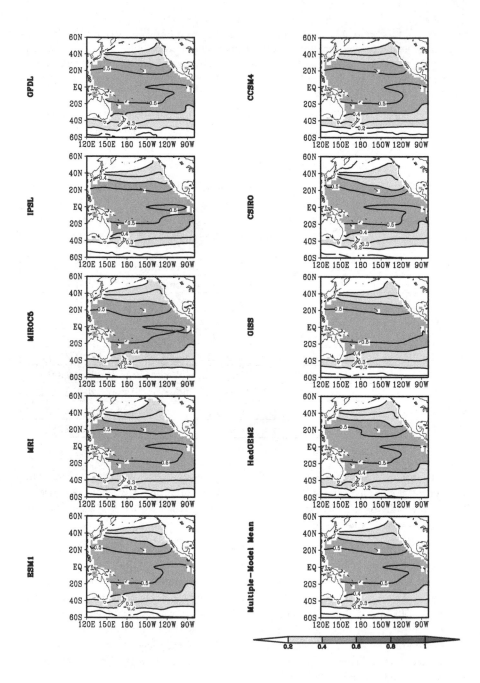

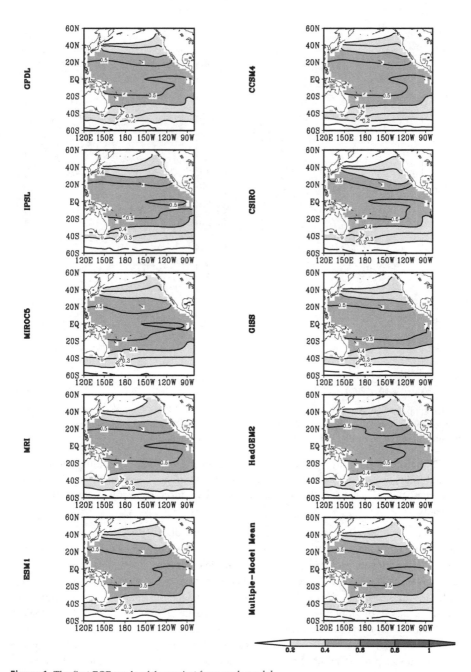

Figure 1. The first EOF mode of the project from each model.

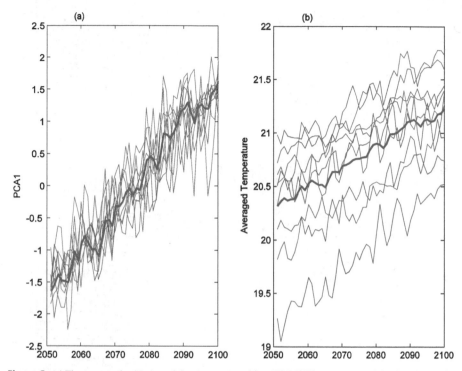

Figure 2. (a) The temporal variation of the time series of first EOF (99% variance) and (b) the averaged temperature over the north America. The blue line is for each model and red line is the mean of all models. The PCA1 of each model in Fig. 2a is normalized prior to plotting.

Fig. 2 shows a visible divergence of projections among models, suggesting uncertainties existed in the responses of these models to the GHGs forcing. It should be noted that little divergence in Fig 2a is due to the normalization, a post-processing just for a good-looking of this figure.

It is of great interest to explore the uncertainty of the above projections. As introduced aforementioned, one can use the above information-based framework to measure the uncertainty of climate prediction, given the multiple ensembles available. Apparently, there are several challenges here: 1) there is only one-member projection for each model, lacking sufficient ensembles; 2) the projection is not dependent on initial condition, thus any measures based on multiple initial conditions are invalid here; 3) the climatological distribution used in estimating the uncertainty may be uncertain under the background of global warming. For the first issue, we propose to solve it using multiple model strategy, i.e. pool all model projections to construct a 9-member ensemble. Under the framework of potential predictability, the model is assumed to be perfect. Thus the disparities among these model projections can be viewed as the ensembles of a perfect model, perturbed by initial conditions or other parameters. For the second issue, we assume that the projection is a long-term prediction at a given initial condition. The distribution for the average of multiple model projections is used as climatological distribution here.

Displayed in Fig. 3 are the variations of projection utility RE during the projection time from 2051 to 2010.The climatological mean and variance are estimated from all 'ensemble' members and years (sample size is 50*9) as in [25]. The projection mean and variance are estimated each year from the 9-member ensemble. As can be seen, it is apparent that the utility R continues to decrease until around 2070 and then bounce after 2080. For the projection during 2070 - 2080, RE is small. When projection (prediction) and climatology distributions are identical, the relative entropy R is zero from (14). In theory, a nonzero value of R indicates predictability. However, in practice, a finite sample size introduces sampling errors that lead to a nonzero R even though there is no extra information supplied by the prediction. Therefore the statistical significance level should exceed the extent of uncertainty due to the finite sample size. We quantify the extent of uncertainty using a Monte Carlo method as in [26]. A sample with 9 members is randomly drawn from the climatology distribution and its relative entropy R is computed with respect to the climatology distribution. This process is repeated 10 000 times, and the value above 95% of 10 000 RE is considered to be the significant level as shown in Fig. 3 (dashed line). As can be seen, the projections between around 2070 and 2080 have statistically 'zero' relative entropy, and the other projections beyond this period have significant relative entropy.

A striking feature of RE in Fig. 3 is its U-shape variation with the projection time (i.e., the time step of integration), which is quite different from actual ensemble forecast at time scales from days to seasons. Typically, the RE monotonously decreases with the lead time of prediction at the time scales from days to seasons (e.g., [1-2, 11]), i.e., the predictability decreases with lead time. The monotonous variation of RE with lead time of predictions well characterizes the nature and attributes of realistic atmospheric and oceanic system, which is chaotic and stochastic, leading to the information at initial conditions gradually dissipated with lead time. Apparently it is not this case here, since the projection is not an initial value problem, and mainly is a response to external forcing (e.g., CO_2).

One possible explanation for this U-shape is related to the climatological distribution used here. We used the average of multiple model projections that have an apparent trend as the climatology distribution. If the RE is dominated by the ensemble mean (ensemble mean square) and the contribution of ensemble spread is relatively much smaller, the RE can show such a U-shape structure. Another plausible explanation is based on a hypothesis, namely, the climatology from multiple models is close to the true value. Under this assumption, the projection with small RE in figure 3 has high fidelity and vice versa. Here, we use the RE to measure the difference between the distribution of projection and the designed distribution, which has been also used in previous studies [17]. However such a hypothesis may cause concerns. One may argue to use present climatological distribution as a reference distribution in the above discussions. However, it can be expected that the climatology of the scenario of R60 should be quite different from the present one. Thus, a further study on the reference distribution is highly demanded in estimating uncertainty of climate change projection.

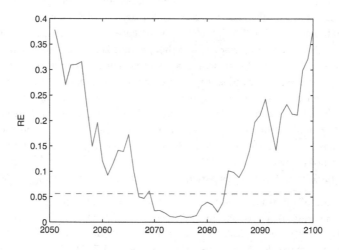

Figure 3. RE as a function of projection time

6. Conclusion

In this chapter, the SNR-based and information-based measures of potential predictability were introduced. They include the signal to noise ratio (SNR) and two measures of information-based predictability. One is relative entropy (RE) that measures individual potential predictability whereas the other is mutual information (MI), the average of RE over all initial conditions, which measures the average potential predictability. From statistical derivation and theoretical analysis, we have below conclusions:

i. The SNR usually measures the average predictability with the assumption that signal inherent to slowly varying external forcing is predictable and the noise is unpredictable;

ii.   A new measure of prediction utility that is derived from information theory is introduced. It measures the additional information provided by a prediction (p) over that already available from the climatological or reference distribution (q). One natural measure is their relative entropy RE defined as the relative difference of entropy between p and q. For the case of Gaussian distributed p and q, the RE can be expressed in terms of the prediction and reference means and covariance.

iii. Averaged RE over all initial conditions, called the mutual information (MI), a measure of the statistical dependence of the forecast state and the initial (boundary) conditions, measure the averaged predictability. The MI-based metrics can measure more potential prediction utility than the SNR-based counterpart. The MI-based predictability measures the statistical dependence, linear or nonlinear, between ensemble mean (prediction) and an ensemble member (hypothetical observation), whereas the SNR-based predictability only measures a linear relationship between prediction and hypothetical observation.

iv.   When the prediction and climatological distribution are Gaussian and the ensemble spread is constant with predictions, both measures are identical to each other. When the ensemble spread is not constant, the SNR-based predictability often underestimates the potential predictability.

v.   The predictable component analysis (PrCA), a method that assesses the most predictable patterns, is introduced. The PrCA decomposes the predictability into patterns accounting for different fractions of the total predictability. Distinguishing spatial structures that are unpredictable from those that are predictable is important for practical prediction problems, particularly when the predictable patterns are few in number.

As an example, the uncertainty of the climate change projection from scenario R60 of AR5 was evaluated, with the Pacific SST as the target. Nine models from different countries were participated in this evaluation. It was found that the most striking warming occurs at the tropical Pacific along the equator. In the extra-tropics beyond 30S to 30N, the increase in SST is relatively weaker. On average, the mean temperature of the Pacific ocean of 60S to 60N increases around 0.5 to 1C from 2051-2010 in these models. The relative entropy RE, measuring the utility of climate projection, continues to decrease until around 2070 and then bounce after 2080. For the projection during 2070 - 2080, RE is small. Under the assumption that the climatology from multiple models is close to the true value, the projection during the period with small RE suggests high fidelity and vice versa.

Author details

Youmin Tang
Environmental Science and Engineering, University of Northern British, Columbia, Prince George, Canada
State Key Laboratory of Satellite Ocean Environment Dynamics, Second Institute of Oceanography, State Oceanic Administration, Hangzhou, P. R. China

Dake Chen
State Key Laboratory of Satellite Ocean Environment Dynamics, Second Institute of Oceanography, State Oceanic Administration, Hangzhou, P. R. China
Lamont-Doherty Earth Observatory, Columbia University, Palisades, NY, USA

Dejian Yang
School of Atmospheric Sciences, Nanjing University

Tao Lian
State Key Laboratory of Satellite Ocean Environment Dynamics, Second Institute of Oceanography, State Oceanic Administration, Hangzhou, P. R. China

Acknowledgement

This work is supported by grants from Canadian NSERC Discovery Program and Chinese NSF 41276029 D0601. This work is also supported by grants from the National Basic Research Program (2013CB430300), the National Science Foundation (91128204) and the State Oceanic Administration (201105018, 151053) of China.

7. References

[1]   Tang, Y., Kleeman, R., and Moore, A. M. Comparison of information-based measures of forecast uncertainty in ensemble ENSO prediction. J. Clim. 2008a; 21, 230–247.

[2]   Tang, Y., Lin, H. and Moore, A. M. Measuring the potential predictability of ensemble climate predictions. J. Geophys. Res. 2008b; 113, D04108, doi:10.1029/2007JD008804.

[3]   Shukla, J. Predictability in the midst of chaos: A scientific basis for climate forecasting. Science, 1998; 282, 728–731.

[4]   Kumar, A., Barnston, A. G., Peng, P., Hoerling, M. P., and Goddard, L.Changes in the spread of the variability of the seasonal mean atmospheric states associated with ENSO. J. Clim. 2000;13, 3139-3151.

[5]   Peng, P., Kumar, A., Wang, W. An analysis of seasonal predictability in coupled model forecasts. Clim. Dyn.2009; 36, 637-648.

[6]   Anderson, J. L., and W. F. Stern, 1996: Evaluating the potential predictive utility of ensemble forecasts. J. Climate, 9, 260–269.

[7]   Sardeshmukh, P. D., Compo, G. P., and Penland C. Changes of probability associated with El Niño. J. Climate, 2000; 13, 4268–4286.

[8]   Cover, T. M., and Thomas, J.A. Elements of Information Theory, pp. 576, Wiley;1991.

[9]   DelSole, T. Predictability and information theory. Part I: Measures of predictability. J. Atmos. Sci. 2004; 61, 2425–2440.

[10] Schneider, T., and Griffies, S. A conceptual framework for predictability studies. J. Clim.1999; 12, 3133–3155.

[11] Tang, Y., Kleeman, R. and Moore, A. M. On the reliability of ENSO dynamical predictions. J. Atmos. Sci.2005; 62, 1770–1791.

[12] Kleeman, R. Measuring dynamical prediction utility using relative entropy. J. Atmos. Sci.2002; 59, 2057–2072.

[13] DelSole, T., and Tippett M. K. Predictability: Recent insights from information theory. Rev. Geophys.2007; 45, RG4002, doi:10.1029/2006RG000202.

[14] Joe, H. Relative entropy measures of multivariate dependence, *J. Amer. Stat. Assoc., 1989: 84,* 157–164.

[15] DelSole, T. Predictability and information theory. Part II: Imperfect forecasts. J. Atmos. Sci.2005; 62, 3368–3381.

[16] Kumar, A. and Hoerling, M.P., 1998. 'Annual cycle of Pacific/North American seasonal predictability associated with different phases of ENSO', J. Climate, 11, 3295-3308.

[17] Shukla, J., DelSole,T. Fennessy, M. Kinter, J. and Paolino, D. Climate Model Fidelity and Projections of Climate Change. Geophys. Res. Lett.2006; 33, L07702, doi:10.1029/2005GL025579.

[18] Yang, D. Tang, Y., Zhang, Y. and Yang, X. Information-Based Potential Predictability of the Asian Summer Monsoon in a Coupled Model. J. Geophys. Res.2012; doi:10.1029/2011JD016775.

[19] Tippett, M. K., Barnston,A.G. and DelSole, T. Comments on "finite samples and uncertainty estimates for skill measures for seasonal prediction". Mon. Weather Rev. 2010; 138, 1487-1493.

[20] Venzke S, Allen MR, Sutton RT, Rowell DP. The atmospheric response over the North Atlantic to decadal changes in sea surface temperatures. J Clim 1999; 12: 2562 – 2584

[21] Allen and Smith 1997 Optimal filtering in singular spectrumanalysis. *Phys. Lett.* 234:419–428

[22] Fukunag. Introduction to statistical pattern recognition, Second Ed., Academic press, Boston. 1990.

[23] Sutton RT, Jewson SP, Rowell DP. 2000. The elements of climate variability in the tropical Atlantic region. *J. Climate* 13:3261–3284

[24] Delsole, T. and Shukla, J. Linear prediction of the Indian monsoon rainfall. Centre for Ocean–Land–Atmosphere Studies (COLA) Technical Report, CTR 114, 2002, p. 58.

[25] Tang, Y., Lin, H. Derome, J. and Tippett, M. K. A predictability measure applied to seasonal predictions of the Arctic Oscillation. J. Clim.2007; 20, 4733–4750.

[26] Tippett, M. K., Kleeman, R. and Tang Y. Measuring the potential utility of seasonal climate predictions. Geophys. Res. Lett. 2004; 31, L22201, doi:10.1029/2004GL021575.

Chapter 7

Effect of Climate Change on Mountain Pine Distribution in Western Tatra Mountains

Jaroslav Solár

Additional information is available at the end of the chapter

1. Introduction

1.1. Climate change in the high mountains

The world is experiencing a period of climate change, which is very frequently discussed on both local and global levels. The growth in the global mean surface temperature by 0,74 °C ± 0,18 °C, over the last 100 years (1906-2005) is probably related to greenhouse gas emissions and further warming will cause many changes in the global climate system during the 21st century (IPCC, 2007). These changes will affect both the abiotic and biotic conditions of the environment.

High mountains ecosystems represent unique areas for the detection of climate change and the assessment of climate-related impacts (Beniston, 2003). Climate change associated with global warming at higher elevations is more pronounced than at low elevations (Beniston & Rebetez, 1996; Giorgi et al., 1997; Diaz & Bradley, 1997). The effect of elevation on surface warming is especially marked in the winter and spring seasons, since it is mostly associated with a decrease in snowpack and is thus enhanced by the snow–albedo feedback (Giorgi et al., 1997). The main ecological driving force is climate, with temperature and the duration of the snow cover as key factors (Gottfried et al., 1999). Changes in air temperature can extend the length of the average annual growing season (Menzel & Fabian, 1999) and can also cause a shift in phenology (Parmesan & Yohe, 2003; Visser & Both, 2005).

1.2. Climate change effect on high mountain ecosystems

Climate change is an important driving force on natural systems (Parmesan & Yohe, 2003). Many studies show that high mountain ecosystems are vulnerable to climate change (e.g. Theurillat & Guisan, 2001; Dullinger et al., 2003a, 2003b; Dirnböck et al., 2011). Global climate change resulting in warmer climate may cause a variety of risks to mountain habitats (Beniston, 2003). Climate change mainly affects the distribution of plant and animal communities (Beckage et al., 2008) and under the expected climate scenarios in the final perspective results in the loss of rare species of alpine habitats (Dirnböck et al., 2011). In the global meta-analyses from Parmesan & Yohe (2003) and Root et al. (2003) significant range shifts toward the poles or toward higher altitudes for many organisms were documented. Large part of these changes may be attributed to increased global temperatures.

In general terms, we expected that climate related changes in mountain ecosystems will be most pronounced in the "ecoclines" (boundary ecosystem), or Ecotones (Theurillat & Guisan, 2001). Distribution of endemic mountain species is typically severely restricted as a spatial response in mountain areas, however, because of mountain topography (Huntley & Baxter, 2002) and, often, the availability of suitable soils (Theurillat et al., 1998). The upper forest limit is commonly referred to as tree line. Timberline or forest line represents one of the most obvious vegetation boundaries (ecoclines). In reality the transition from the uppermost closed montane forests to the treeless alpine vegetation is commonly not a line, but a steep gradient of increasing stand fragmentation and stuntedness, often called the tree line ecotone or the tree line park land (Körner & Paulsen, 2004).

1.3. Climate change-induced shift of ecosystem boundaries

Scenarios of upward plant species and vegetation shifts are widely discussed in many current research articles. Theurillat & Guisan (2001) released a review discussing this matter concluding that although the alpine vegetation can tolerate an increase of 1-2 °C of average air temperature, in the case of a sharper increase we can expect major changes. Loss of diversity of the alpine communities and fragmentation of plant populations caused by climate warming is expected for comparable high mountains around the world (Grabherr et al, 1995; Sætersdal et al., 1998).

Results from Dirnböck et al. (2003) support the hypothesis, that alpine plant species above the forest line will be affected by heavy fragmentation and habitat loss, but only if the average annual temperature increases by 2 °C or more. Most of these lost alpine plant species habitats are expected to be caused by the expansion of *P. mugo* in the Alpine zone. The coniferous forest zone has a general tendency to expand to higher elevations (Mihai et al., 2007; Sitko & Troll, 2008). Nicolussi & Patzelt (2006) describe the alpine timberline zone as very sensitive to climate variability. The rise of temperatures during the vegetation period over long periods also induces a rise of the tree line, with higher forest stand density. The tree line is considered to be primarily by temperature controlled, so increases in temperature should result in their upslope expansion (Moen, 2006). Growth and fertility of *Pinus mugo* is mostly controlled by temperature (Dullinger et al., 2004). Thus the main limiting factor of *Pinus mugo* growth at high altitude could be the soil temperature (Smith et al., 2003) although Rossi et al. (2007) refers to varying soil temperature thresholds at different sites, indicating that soil temperature may not be the main factor limiting xylogenesis of conifers and provides strong evidence that air temperature is a critical factor limiting xylem cell production and differentiation at high altitudes. However, the air temperature alone may not be the dominant factor determining tree line position, because the direct influence of temperature may be masked by interactions with other factors such as precipitation, cold-induced photoinhibition, disturbance or plant - plant interactions (Harsch, 2009). This evidence is therefore in conclusive. Differences in expert opinions on this matter have lead Smith et al. (2009) to formulate six current hypotheses of the causes of upper tree limit movement: climatic stress, mechanical disturbance, insufficient carbon balance, limitations of the cell growth and tissue formation, limited nutrient supply, and limited regeneration. In the global meta-analyses from Parmesan & Yohe (2003) and Root et al. (2003) significant range shifts toward the poles or toward higher altitudes for many organisms were documented. Large part of these changes may be attributed to increased global temperatures. The expansion of tree line forming species (sub-alpine zone) to higher altitudes is evident in the Pyrenees (Camarero & Gutiérrez, 2004; Peñuelas et al., 2007), in the Alps (Dullinger et al., 2003a, 2003b; Gehrig-Fasel et al., 2007; Vittoz et al., 2008), in the Carpathians (Martazinova et al., 2009; Mihai et al. 2007; Švajda et al., 2011), Sweden (Kullman, 2002), Caucasus (Akatov, 2009) but also in Patagonia (Daniels & Veblen, 2004) and Himalaya (Song et al., 2004; Becker et al., 2007).

Lapin et al. (2005) detected climate changes in the Slovak mountains. The results showed a significant increase in temperature and a decrease in relative humidity in the April to August season after 1990. From 1901–2005, air temperature increased (annual mean) moderately and precipitation decreased (Melo, 2007). This trend of warming is expected to continue in the Slovak mountains. By 2075 the annual average air temperature in Slovakia is expected to increase by 2-4 ° C (with greater warming expected in the winter) and more significant effects of increasing temperatures is expected at the higher altitudes (Mindáš & Škvarenina, 2003). Generally, climate conditions and land use in high mountain areas have been shown to influence the distribution of mountain pine. Since 1965 the ban on grazing in the High Tatras has not yet been raised. This study assesses a potential scenario after the grazing in Tatra Mountains will have been resumed. The potential model of timberline is based on the assumption that climate change as a factor in forest regeneration is primarily responsible for moving the upper limit of the natural forest above the original climatically determined timberline, while the abandonment of farming in the country should be the dominant factor determining forest regeneration below this line.

2. Study area

2.1. Description of study area

The Tatra Mountains are situated at the Slovak–Polish border (20°10′E, 49°10′N) and constitute the highest mountain massif within the Carpathian Range of Central Europe. The highest summit reaches 2656 m; the massif is classified as a high-mountain landscape covered by subalpine and alpine zones.

The study area (Figure 1) is situated in the western Tatra Mountains. The geology of the investigated area is based on crystalline bedrock. The western Tatra Mountains contain a significant amount of metamorphics (gneiss and mica schist), in addition to granodiorite

(Nemčok et al., 1993). The vegetation of the alpine zone is dominated by alpine meadows (dry tundra with mostly *Festuca picturata, Luzula alpino-pilosa, Calamagrostis villosa,* and *Juncus trifidus*), with patches of dwarf pine (*Pinus mugo*) and an increasing percentage of rocks (bare or covered with lichens—commonly *Rhizocarpon, Acarospora oxytona,* and *Dermatocarpon luridum*) above the upper tree line of 1800 masl (Vološčuk, 1994).

The average annual air temperature decreases with elevation by 0.6°C per 100 m, being 1.6 and 23.8°C at elevations of 1778 and 2635 m, respectively (Konček & Orlicz, 1974). The amount of precipitation increases with elevation, varying from ~1.0 to ~1.6 m yr^{-1} between 1330 and 2635 masl but reaching >2.00 m yr^{-1} 21 in some valleys (Chomitz & Šamaj, 1974). Precipitation is generally higher in the northern part than in the southern part of the mountains, as is runoff, which averages 1.42 and 1.57 m yr^{-1} for the south and north, respectively (Lajczak, 1996). Snow cover usually lasts from October to June at elevations > 2000 masl.

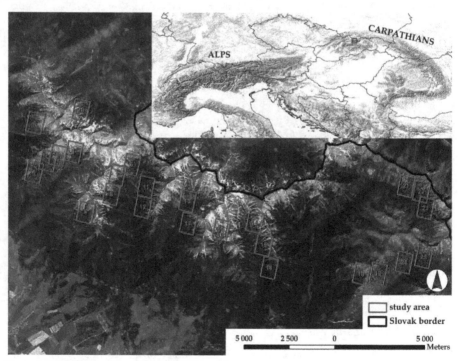

Figure 1. Western Tatra Mountains in Slovakia and detailed view of the 25 sites in the study area (from west to east: Roháče, Baníkov, Baranec, Bystrá, Jamnická, Račkova, Kamenistá, Tichá, Kôprová, and Špania valleys). (Map by Jaroslav Solár)

The relationship between climate conditions of the environment (microclimate and vertical climate) and phytocenoses is expressed at different altitudinal zones in the forest. Constant climate conditions definitively influenced the natural distribution of forest species from the sub-Atlantic period (around 2000 years ago), when the current altitudinal zones were formed. Significant changes of forest stratification were caused by the intense human activity since the 13th century. Ecologically, forest altitudinal zones represent vertical classification of vegetation. Horizontal classification is determined by growth condition of forest societies, differentiated especially according to soil conditions, ecological rows, interrows, and hydric files of forest type groups. The climate-driven tree line in the Tatra Mountains is located around 1550 masl and partly includes natural ecotones with individual conifers reaching ages of 350–450 years (Büntgen et al., 2007).

P. mugo is an obligatory prostrate pine with adult canopy height varying between 0.3 and 2.5 m in the study area. The typical dwarf pine altitudinal (subalpine) zone extends from 1500 to between 1850 and 1900 masl. Mountain pine zone developed especially in the western Tatras with glacial-meadow relief, with great antierosion and water retention potential. Closed mountain-pine thickets stretch up to 300 m above the timberline, reaching approximately 1600–1750 masl in the Tatras and encompassing the upper part of the forest alpine tundra ecotone. Mountain pine plays a significant role in the natural environment: it protects the soil and stabilizes the snow cover, thus restricting the release of avalanches, and it provides habitat for many species of flora and fauna (Jodłowski, 2006).

2.2. Climate change in study area

On slovakia in the period 1881-2007 was increase of annual temperature in 1,6°C and annual precipitation decrease in 24 mm (Lapin et al., 2009). The temperature series show an upward trend in all seasons, especially in the spring (Melo et al., 2009). Over the past 20 years, it seems much warmer and especially in the months of January to August (Faško et al., 2008). Warming scenarios based on applied GCM (General Circulation Model) for Slovakia represent the increase in average annual temperature of 2-4 °C until the end of the 21st century (Melo et al., 2009). The climate in the Slovak mountain region is thus becoming warmer. Figure 2 shows a general trend that could partially explain the dynamics of the vegetation zones.

Winter precipitation in the high-mountain positions of Slovakia is abundant and increases with altitude. (Ostrožlík, 2008, 2010). Sensitivity of snow cover will vary depending on the climate and altitude. Also sensitivity causes maritime climates and less continental climate of cold and dry winters, where precipitations play an important role in the variability of snow cover duration. (Brown & Mote, 2009). The Tatra Mountain have significanly more snow cover days on the northern slopes. More over the less windy and forested areas have higher and longer snow cover as well. (Lapin et al., 2007). Snow cover duration on the northern slopes is critical in altitude of about 1800 m and on the southern slopes of about 2300 m. In this altitudinal level in Tatra mountain should be zone, above which would not even occur to the loss of snow cover duration. (Vojtek et al., 2003). It seems that variability and trends in snow cover characteristics are influenced both by air temperature and precipitation variability. This influence depends significantly on the altitude and local topography conditions. Increase of air temperature by about 1,2 °C and change of precipitation totals from -10% to -20% in the November-April season are the main reasons of obtained trends (Lapin et al., 2007).

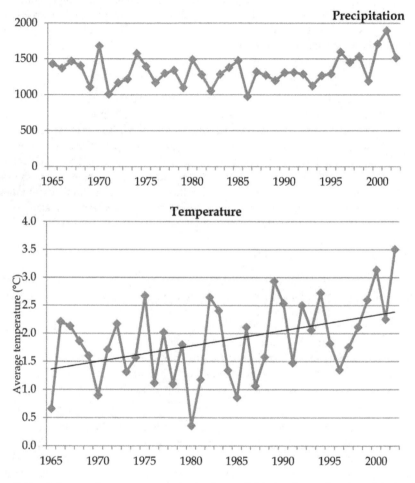

Figure 2. Trends in annual average temperature (in degrees Celsius) and annual total precipitation (in millimeters) from 1965 to 2002 at the meteorological station of Skalnate´ pleso (1751 masl), Slovak Institute of Hydrometeorology (Švajda et al., 2011).

3. Material and research methods

3.1. Theoretical aspects

Changes in landscape can be well observed through remote sensing (RS). RS data (images, aerial photographs, etc.) are further processed and analyzed using Geographic Information

Systems (GIS). Progressive development of geo-information technologies offer new approaches to the use of remote sensing in GIS. GIS is very useful for its ability to incorporate the complexity of spatial data in to the various models. Remote-sensing based analysis is particularly useful in mountainous areas where the topography is complex and different environmental gradients require special attention to the spatial patterns (Heywood et al., 1994). Although high mountain environments show a high degree of heterogeneity, we can obtain satisfactory results using the appropriate approaches and high-quality materials. Changes in the natural spatial (morphological, bioenergetical) features can be identified in remote sensing images (Feranec et al., 1997). Using GIS applied approach let us identify the most significant variables and phenomena which affect the natural environment components.

Approach of remote sensing and suitability of this method in detecting of lanscape changes in mountain regions of Slovakia was confirmed in the works: Boltižiar, (2001, 2002, 2003, 2004, 2006, 2007); Čerňanský & Kožuch, (2001); Hreško & Boltižiar, (2001); Kohút, (2006); Olah et al., (2006); Falťan & Saksa, (2007); Olah & Boltižiar, (2009). However, this approach can be difficult in countries with politically sensitive situation, where the products of remote sensing are subject to various degrees of secrecy (Heywood et al., 1994). We have come across some other problems arise in relation to data quality and its precision of position placement, which takes into account the high diversity of the relief. Advantage of access interpolation of aerial imagery lies in the fact, that we can carry out their research in a relatively short time. Especially aerial photographs provides a large amount of quantitative and qualitative information about the landscape structure and are particularly important in high mountain areas, where field research is difficult (Boltižiar, 2009).

This study is based on results published in the original study by Švajda, Solár, Janiga & Buliak „Dwarf Pine (Pinus mugo) and Selected Abiotic Habitat Conditions in the Western Tatra Mountains" in journal Mountain Research and Development 31/3 in year 2011. The present analysis was carried out using GIS (ArcGIS 9.3), based on aerial photographs from 1965, 1986 and 2003. The aim of the analysis was to verify the temporal trends in the distribution of Pinus mugo and to investigate which environmental variables best explain the changes in the growth and distribution of the mountain pine. The applied modelling approach is based on three major assumptions: (1) The abiotic factors are assumed to be the major driving force of species distribution changes, as well as the post-grazing succession. (2) The models are calibrated using field data, and thus comprise any competitive constraint a species may force upon or experience from its neighbour. (3) The speed of plant migration is consistent with that of climate change so that plant communities are in a permanent equilibrium with their environment (Dirnböck et al., 2003).

3.2. Materials and data processing

Aerial images from 2002 were acquired and georeferenced by Eurosense Ltd. and Geodis Slovakia on the basis of contour lines at a scale 1:10,000 (digital elevation model); we georeferenced aerial photographs from 1986 and 1965 on the basis of orthophotos (Table 1).

Year	Source	Type	Resolution	Width	Height	Format
2002	Eurosense Slovakia	Orthophoto	RBG 72 DPI	2500	2000	.jpg
1986	Topographical Institute	Aerial photo	Gray 2400 DPI	21.829	21.924	.tiff
1965	Knazovicky	Aerial photo	Gray 300 DPI	2164	2175	.jpg
DPI. dots per inch; RBG. red green blue.						

Table 1. Quality and resolution of data features (aerial imagery) (Švajda et al., 2011).

Mountain pine fields were extracted from the aerial photos in gray scale and than reclassified into gray scale range representing mountain pine occurrence in the study area. Each photo was examined individually. If mountain pine on the slide was gray, with a value from 75 to 110, all such values in the range were reclassified as 1. The remaining values from 0 to 75 and 110 to 256 were reclassified as 0. We created a grid where each pixel contained either the value 1 or the value 0. Then the grid was automatically vectorized on the basis of the 2 values.

Habitat conditions were spatially simulated using GIS, digital terrain model, meteorological data and existing maps. In addition we analyzed historical records in order to derive information about past land-use changes. The most significant factors explaining the presence of *Pinus mugo* according to Dirnböck et al., (2003) are the daily temperature, followed by slope, geology, solar radiation in September and duration of snow cover.

To test this hypothesis it was necessary to create an explicit temporal and spatial explicit model of the spread of mountain pine and analyze their sensitivity to predicted climate change trends. Histogram transformation was not carried out due to the misrepresentation of values. The size of the pixels' (cell) grid was equivalentin all RS images, because each image was adjusted to the same size cell size through the transformation of the grid, as well as during georeferencing. Thus all images and the grids had the same pixels (Figure 3).

Selection of the appropriate areas, which represented 25 localities from the study area, and analysis of imagery were realized in ArcGIS 9.2. Differences in the *P. mugo* surface cover between the 2 periods were calculated using Statistica 8.

The increments in dwarf pine were reported as means and standard deviations for potential comparison with other studies, but the values showed a highly skewed distribution in most sample groups. Therefore, a nonparametric approach to the analysis of the data was necessary. The significance of difference between groups was tested using the Kruskal–Wallis nonparametric test. When P , 0.05, the data were considered as significantly different. A digital elevation model of the study area was used for the representation of a selected abiotic habitat conditions. A single matrix was analyzed. GIS intersection of study sites with 3 parameters (slope, aspect, and height masl) has divided the studied sites into 325 smaller areas with unique characteristics related to pine increase. Two sites (nos. 5 and 9; Table 2) were excluded due to lack of data from 1986.

The principal component analysis (PCA)–correlation matrix, a multivariate technique was used to extract the potential relationships between the studied variables. Principal components are linear combinations of original variables (slope, height masl, and relative increase of pine during observed periods), each axis being statistically orthogonal to the others. Integration of the variables slope and elevation m asl in different periods enabled us to follow different processes of mountain colonization by mountain pine during the respective periods. Since this statisticale technique produces statistically orthogonal axes, we were able to examine potentially independent biological phenomena. We used 4 variables; consequently, we evaluated 4 principal components. The proportions of the total variance accounted for by each component are shown in Table 3 (see results).

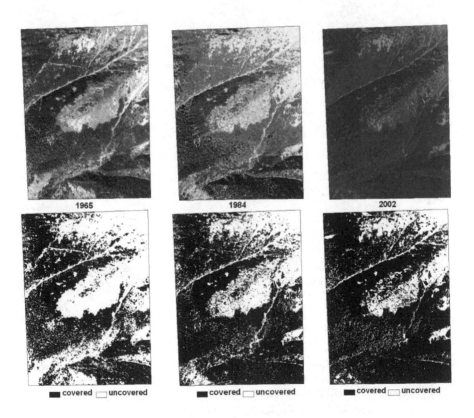

Figure 3. Example of the comparison between aerial photographs: changes between 1965, 1986, and 2002 in 1 analyzed valley (Site 17, Račková valley; Figure 1).

Site number	Average altitude (masl)	Average slope (%)	Covered with *Plnus* (%)	Difference (+/-)
1	1674	34	27	+1/+20
2	1670	46	7	+9/+11
3	1614	44	29	+9/+23
4	1675	36	24	+31/+41
5 (excluded)	1686	42	—	—
6	1614	33	35	+8/+-10
7	1733	28	28	+4/+9
8	1604	40	53	+11/+13
9 (excluded)	1807	34	—	—
10	1825	22	16	+6/+16
11	1511	47	61	+14/+14
12	1533	43	68	+12A16
13	1493	33	47	+8/+10
14	1684	40	19	+4/+12
15	1617	46	21	+11/+13
16	1561	43	62	+3/+10
17	1640	45	48	+18/+28
18	1777	35	34	+6/+11
19	1595	55	45	+10/+12
20	1627	30	42	+6/+23
21	1449	33	49	+21/+20
22	1632	45	42	+11/+31
23	1599	36	68	+11/+20
24	1428	31	59	+20/+16
25	1533	38	65	-2/+2

Table 2. Overview of evaluated sites with different *P. mugo* cover in the period.

4. Results

4.1. Surface cover of mountain pine

Mountain pine cover in the western Tatra Mountains in the period 1965–2002 permanently increased at all observed sites. The total surface area covered by mountain pine increased from 8,173,812 m^2 in 1965 to 10,141,505 m^2 in 1986 and 11,394,461 m^2 in 2002. The percentage of total surface area covered thus increased from 41.8% in 1965 to 51.8% in 1986 and 58.2% in 2002. Only in one case (No. 25) surface area covered by dwarf pine decreased. In two cases (No 21 and 24) the area decreased in the first, and increased in the second period (Table 2, Figure 4). This was probably due to the influence by human activities or avalanches.

The results also indicate that the mean increase of mountain pine surface cover was in all periods about 0.4 percent per year (0.42% first period, 0.40% second period) from the total surface area but results in relation to selected abiotic conditions still showed some differences.

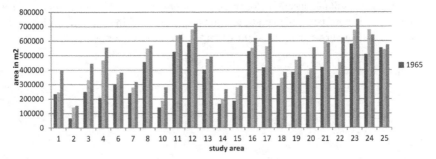

Figure 4. Comparison of area covered with *P. mugo* (23 sites) in 1965, 1986, and 2002, in square meters (Švajda et al., 2011).

4.2. Expansion of mountain pine growth

From 1965 to 1986, mountain pine showed a rapid expansion in surface cover at the lower elevations (Table 3 – PC1, Figure 5A). This could be observed as thickening of mountain pine cover at lower elevations, indicating that the mountain pine is able to recolonize sites of previous occurrence.

Variable	PCI	PC2	PC3	PC4
Slope	0.51	-0.64	-0.50	0.26
Elevation	-0.71	0.11	-0.63	-0.27
Pine increment (1965-1986)	0.73	0.14	-0.10	-0.64
Pine increment (1986-2002)	-0.38	-0.78	0.30	-0.37
Variability (%)	36.5	26.8	18.9	17.8

PC. principal component.

Table 3. Component vectors (loadings) and percent variance associated with the components indicating the pattern of natural reforestation with dwarf pine in the Tatra Mountains (n 5 325; snaps from aerial photographs) (Švajda et al., 2011).

In the period from 1986 to 2002, pine grew rapidly on steeper slopes (Table 3 – PC2), mainly at elevations from 1500 to 1700 masl (Figure 5B). During the first analyzed period, mountain pine was able to concentrate to such a level at elevations between 1300–1400 m a.s.l. that in the following period it completely covered this zone and further increments were minimal.

The third factor (Table 3 - PC3) is less important for the explanation of the historical pine increments: it shows a positive relation between slope and elevation. PC4 (Table 3) is a unipolar vector indicating that mountain pine grew more rapidly in the earlier period (1965–1986) than later (1986–2002).

In the earlier period mountain pine grew intensely at all locations (Figure 5C) whereas in the period 1986–2002 it mainly preferred northwest and northeast aspects on steeper slopes, probably the most suitable locations for plant development in the Tatra Mountains. These sites might have more favorable conditions for the growth of the mountain pine due to changes in climate in terms of higher surface temperature of environment.

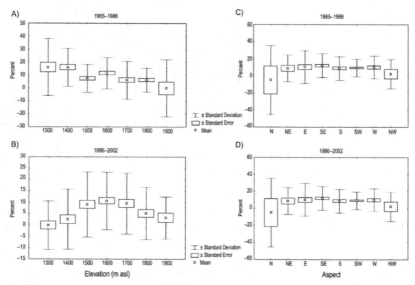

Figure 5. Increments in dwarf pine cover in the western Tatra Mountains in the periods 1965–1986 and 1986–2002, according to (A and B) elevation and (C and D) aspect. In the earlier period, the groups did not differ according to aspect. (C) Kruskal–Wallis nonparametric test at p = 0.05. (D) In the period 1986–2002, the following aspects differed significantly: N:NE, N:SW, NE:W, NE:S, NE:SE, SE:SW, S:SW, and SW:W (Švajda et al., 2011).

In both periods, the increments in the areas covered by mountain pine were very low at the elevation of 1900 m, reflecting its natural upper line of occurrence. In the earlier monitored period the increments at 1900 m were 0%, whereas between 1986 and 2002 they were approximately 3% (Figures 5A, 4B). The trend is probably associated with climate warming in the region. Changed enviromental conditions caused by climate change promote the expansion of mountain pine and favour it in competition against alpine meadow communities.

5. Discussion

The interaction of individual components of the environment is an ongoing process. The result of this interaction as seen in our results, show an apparent shift of mountain pine to higher altitudes. The expansion of mountain pine was confirmed by remote sensing. The precision of our results was limited by the fact that we performed a very fast automatic extraction of mountain pine fields. However, this analysis was repeated several times in order to avoid any errors during the extraction of mountain pine fields. We also recorded other factors which might affect the results of the distribution and growth of dwarf pine. This mainly relates to landslides, avlanches, snow cover and shadows from clouds or hills on the air photographs. Generally, a place where we have identified these problems, we excluded from the assessment in all of times periods. Due to aim of this study was not to highlight the processes that operate in the opposite direction to the expansion of mountain pine, so we did not dealt with this problem more. We had some problems in the lower parts of fields where the scrub of mountain pine interleaved with spruce forest. Analysis of this border and its response to climate change would be also interesting. We can assume, that spruce forest has pushed the lower limit of mountain pine to the higher altitudes (Mihai et al., 2007). But this process is slower than the expansion of mountain pine due to problems with the successful survival of spruce seedlings and seed production (Dullinger et al., 2005). Evidence of spruce forest move to higher altitudes was shown by Mihai et al. (2007) in the Southern Carpathians of Romania. In comparison to spruce stands the mountain pine cover in our study represented comparatively homogeneous areas easy to extract in our aerial images.

Similarly to other high mountains also the Carpathians show a trend of climate change and possibly the shift of vegetation types with altitude. Expected changes in tree line boundaries are evident in Carpathians but also other mountain ranges around the world, which could present a threat to the habitats of many rare species in the future. Over the last 50 years, summer temperatures in the Tatra Mountains summer temperatures have increased by 0.7 ° C at higher elevations, and 1.4 ° C at lower elevations. Winter temperatures have increased by 1.4 ° C at higher elevations, and 1.9 ° C at lower elevations (Melo, 2005). The temperature limit of the mountain pine zone is determined by the bio-temperature threshold in the range of 3.0 to 2.0 ° C (°C Max - Min ° C) (Miňdáš & Škvarenina, 2003). Considering the rate of current temperature changes (2-3 °C for 100 years) we can expect more turmoil changes to the growth within a single generation of woody plants. According to Miňdáš et al. (1996) a model scenario expects a complete extinction of conditions for alpine communities and their replacement by bioclimatic conditions for sub-alpine forest. The occurrence of mountain pine is subject to extreme habitat conditions, including soil. Mountain pine is a strongly heliophilic shrub. The most important factor of habitat which has a decisive role in the expansion of mountain pine is the light intensity. The spreading of mountain pine is conditioned mainly by altitude, slope, moisture conditions, but also the horizontal and vertical slope curvature. This can be seen in the fact that the mountain pine is spreading up along the ridges.

Miňdáš et al. (2004) predict the following changes in an area of mountain pine zone timberline: (1) an increase in the abundance of tree species, (2) dominant representation of spruce, (3) a decrease of dwarf pine, and (4) an increase of general production and biomass of about 200–300%. These changes could also be contributed by the changes in phenological phases, which reflect the changing climate condtions. The onset of individual phenological stages and their proceeding is mainly influenced by air temperature, as well as temperature and humidity of soil and other meteorological variables (Škvareninová, 2009). Development of climate can to some extent affect phenological trends (Bauer 2006; Škvareninová, 2008) and identyfying these relationships can help us use trees as bio-climatic indicators of climate change (Škvareninová, 2009). At high altitudes the vegetation is under constant environmental stress and thus abiotic conditions become more important for the community development than biotic relationships (Pauli et al., 1996).

The main results of our case study confirm the results of previous research on mountain vegetation zones in the Slovak Tatras. Boltižiar (2007) analyzed spatiotemporal landscape structure change in the alpine environment of the Tatra Mountains. The landscape structure in 1949 in the study area was dominated by grassland, which resulted mainly from human activity. Statistical analysis of thematic maps from 2003 suggests extension of mountain pine cover, advance of forest, and reduction of grassland areas. Martazinova et al. (2009) conducted research on grasslands above the upper forest limit in the Ukrainian Carpathians. Grass cover significantly decreased in the sites whit conifer presence. Spruce stands mainly on the northern slopes moved to higher altitudes, while the beech stands in the same area on the southern slopes did not show any significant movement. Apparently the greatest changes were recorded at those sites where upper forest limit was marked at higher elevations. In their study of alpine, subalpine, and forest landscapes in the Iezer Mountains (southern Carpathians), Mihai et al. (2007) described how mountain pine–subalpine associations developed and gradually covered subalpine meadows and barren land (between 1986 and 2002, colonization averaged 0.14 km2/y). This might be important in the context of the surface of the subalpine and alpine zones in the mountains. However, mountain pine area has lost some lower stands because of spruce forests, which increased in elevation. This is largely a feature of southern aspect slopes (sunny), where the natural timberline is under some local conditions higher. It is also related to shorter duration of snow cover on the southern slopes (Lapin et al., 2007). Peneuelas et al., (2007) also observed a shift and change in the distribution of species on the tree line in the Montseny Mountains (Span). As observed from historic photographs for the last 60 years, beech stands significantly increased in abundance, which is reflected in the shift of this species to higher altitudes by about 30-50 meters.

However, there are interesting comparisons with studies from other European mountains. The results of the study conducted by Dirnböck et al. (2003) support earlier hypotheses that alpine plant species on mountain ranges with restricted habitat availability above the tree line will experience severe fragmentation and habitat loss, but only if the mean annual temperature increases by 2 °C or more. Even in temperate alpine regions, it is important to consider precipitation, in addition to temperature, when climate impacts are to be assessed. Another example from the Alps in Austria (Dullinger et al., 2004), after running a model for 1000 years, predicted that the area covered by pines will increase from 10% to between 24% and 59% of the studied landscape. The shape of the dispersal curve and spatial patterns of competitively controlled recruitment suppression affect range size dynamics at least as severely as does variation in assumed future mean annual temperature (between 0 and 2°C above the current mean). Moreover, invasibility and shape of the dispersal curve interacted with each other due to the spatial patterns of vegetation cover in the region. Dullinger et al. (2003a) indicated that a shift of tree and shrub species caused by landuse and expected climate change can be expected in the European Alps. Abandonment of pasture will allow invasive expansion of *Pinus mugo* scrubs to new areas. In the peripheral areas this process will be dependent on the competitive struggle for light with abandoned grasslands after the grazing has ceased. Gehrig-Fasel et al. (2007) compared upward shifts to the potential regional tree line by calculating the difference in elevation of the respective pixels. The altitude of the potential regional tree line was considered as a reference. Upward shifts above the potential regional tree line were considered to be influenced primarily by climate change, while upward shifts below the potential regional tree line were interpreted as primarily influenced by land abandonment. Generally, dwarf pine forest lost a total surface area under pressure from lower vegetation communities and even secondary pastures (Mihai et al., 2007).

In addition to climate change, human land use may drive changes in tree line. Land use in subalpine and alpine areas (grazing and extraction) affects the distribution of flora just as much as climate. Since the 13th–14th century, anthropogenic land cover change has involved clearing mountain-pine thickets to obtain new pastures for sheep and cattle grazing, for extensive charcoal and oil production, and for copper and iron-ore mining, sometimes leading to degradation. Jodłowski (2007) described how establishing national parks in the Tatras—Babia Góra and Giant Mountains enabled secondary succession, which has led to colonization of previously abandoned habitats. However, these processes have been hampered by harsh edaphic and climatic conditions as well as by avalanches and debris flows. Extensive planting of mountain pine in former Czechoslovakia significantly facilitated the regeneration of mountain pine thickets. After the absolute restriction of grazing in some national parks, we observed progressive long-term trends in secondary succession and patterns of plant establishment driven by climate.

Closed mountain pine thickets stretch up to 300 m above timberline, reaching approximately 1600–1750 masl in the Tatras and encompassing the upper part of the forest-alpine tundra ecotone. Habitats in the peripheral or isolated mountain belts at or above the tree line are generally rich in diversity of endemic species. In these habitats, tree line expansion disproportionally reduces habitats of high-altitude species. Such legacies of climate history, which may aggravate extinction risks under future climate change, have to be expected for many temperate mountain ranges (Dirnböck et al., 2011). Minimizing greenhouse gas emissions effectively in order to reduce climate warming, and thus the expansion of tree line species to higher altitudes. Furthermore, slowing down forest expansion by land use. The maintenance of large summer farms may contribute to preventing the expected loss of nonforest habitats for alpine plant species and might provide additional refuges for those endemic species which can survive in managed habitats (Dirnböck et al., 2003).

6. Conclusion

Our study shows an apparent shift and densification of *Pinus mugo* scrubs at higher altitudes. This shift was shown to correlate with climate change. Longer growing seasons, milder winters, shorter duration of snow cover create favourable conditions for the growth of mountain pine. This shift has not only had a devastating effect on alpine plant communities due to habitat loss, but also due to greater fragmentation, which ultimately will strongly affect the population of different animal species dependent on these habitats.

More research on vegetation dynamics in Slovakia's mountain areas is needed in light of the significance of vegetation in the context of global change. The results of our study can be used not only as a baseline for future research to test possible climate change influences (resulting upward shifts compared to a potential surface size and trends in approach of dwarf pine extension) but also to compare trends in other mountainous areas. Further understanding of dispersal, persistence, and survival strategies of mountain pine in the western Carpathians is also required. We will continue to monitor dispersal of *P. mugo* in Slovakia and extend our studies to the central Tatras. This workwill help to describe and evaluate the total tree surface area as a basis for the State Nature Conservancy's management of mountain national parks and protected areas in Slovakia.

Author details

Jaroslav Solár
Institute of High Mountain Biology, University of Zilina, Slovakia

Acknowledgement

This research was partly supported by the European Economic Area and Norwegian financial mechanism grant SK-00061. We thank Dr L. Kňazovický for aerial photos from 1965. Let us also thank to authors Dr. J. Švajda, Prof. M. Janiga and M. Buliak who created the previous study entitled "Dwarf Pine (Pinus mugo) and Selected Habitat Abiotic Conditions in the Western Tatra Mountains" published in journal Mountain Research and Development 31/3 in the year 2011.

7. References

Akatov, P. V. (2009). Changes in the upper limits of tree species distribution in the Western Caucasus (Belaya river basin) related to recent climate warming. *Russian Journal of Ecology*, Vol.40, No.1, pp. 33–38, ISSN 1067-4136

Bauer, Z. (2006). Fenologické tendence složek jihomoravského lesa na příkladu habrojilmové jaseniny (Ulmi-Fraxineta Carpini) za období 1961–2000. Část II. Fenologie bylin a ptáků. *Meteorologické zprávy*, Vol.59, No.3, pp. 113–117, ISSN 0026 - 1173

Beckage, B., Osborne, B., Gavin, D. G., Pucko, C., Siccama, T. & Perkins, T. (2008). A rapid upward shift of a forest ecotone during 40 years of warming in the Green Mountains of Vermont. *Proceedings of the National Academy of Sciences of the USA*, Vol.105, No.11, pp. 4197–4202, ISSN 1091-6490

Becker, A., Körner, C., Brun, J., Guisan, A. & Tappelner, U. (2007). Ecological and land use studies along elevational gradients. *Mountain Research and Development*, Vol.27, No.1, pp. 58–65, ISSN 1994-7151

Beniston, M. & Rebetez, M. (1996). Regional behavior of minimum temperatures in Switzerland for the period 1979–1993. *Theoretical and Applied Climatology*, Vol.53, No.4, pp. 231–243, ISSN 1434-4483

Beniston, M. (2003). Climatic change in mountain regions: a review of possible impacts. *Climatic Change*, Vol.59, No.1-2, pp. 5-31, ISSN 1573-1480

Boltižiar, M. (2001). Evaluation of vulnerability of high-mountain landscape on example Velická valley in the High Tatra Mts. *Ekológia*, Vol.20, No.4, pp. 101-109, ISSN 1377-947X

Boltižiar, M. (2002). Analýza krajinnej štruktúry vysokohorskej krajiny Tatier vo veľkých mierkach v prostredí GIS. In: *Geografické informácie, Vol. 7, No. 1, Zborník z XIII. zjazdu SGS pri SAV*, Dubcová, A. & Kramáreková, H. (Ed.), pp. 288-297, FPV UKF, ISBN 80-8050-542-x, Nitra, Slovakia

Boltižiar, M. (2003). Mapovanie a analýza vzťahu krajinnej štruktúry a reliéfu vysokohorskej krajiny Tatier s využitím výsledkov DPZ a GIS. *Kartografické listy*, Vol.11, pp. 5-15, ISSN 1336-5274

Boltižiar, M. (2004). Analýza zmien krajinnej štruktúry vybranej časti Belianskych Tatier v rokoch 1949-1998 aplikáciou výsledkov DPZ a GIS. In: *Štúdie o Tatranskom národnom parku 7 (40)*, Koreň, M., (Ed.), pp. 483-491, Marmota press, ISBN 8096887858, Poprad, Slovakia

Boltižiar, M. (2006). Changes of high mountain landscape structure in the selected area of Predné Meďodoly valley (Belianske Tatry Mts.) in 1949-1998. *Ekológia*, Vol.22, No.3, pp. 341-348, ISSN 1377-947X

Boltižiar, M. (2007). *High-mountain landscape structure in Tatras : Large-scale mapping, analyses and evaluations of changes by application of data of the Earth remote sensing*, Faculty of Natural Sciences UCP & Institute of landscape ecology SAS & Slovak Committee of the UNESCO programme MAB, ISBN 978-80-8094-197-0, Nitra, Slovakia

Boltižiar, M. (2009). *Vplyv georeliéfu a morfodynamických procesov na priestorové usporiadanie štruktúry vysokohorskej krajiny Tatier*, Univerzita Konštantína Filozofa v Nitre & Ústav krajinnej ekológie SAV, ISBN 978-80-8094-544-2, Nitra, Slovakia

Brown, R. D. & Mote P., W. (2009). The Response of Northern Hemisphere Snow Cover to a Changing Climate. *Journal of Climate*, Vol.22, No.8, pp. 2124-2145, ISSN 1520-0442

Büntgen, U., Frank, D. C., Kaczka, R. J., Verstege, A., Zwijacz-Kozica, T. & Esper, J. (2007). Growth responses to climate in a multi-species tree-ring network in the Western Carpathian Tatra Mountains, Poland and Slovakia. *Tree Physiology*, Vol.27, No.5, pp. 689–702, ISSN 1758–446

Camarero, J. J. & Gutiérrez, E. (2004). Pace and Pattern of Recent Treeline Dynamics: Response of Ecotones to Climatic Variability in the Spanish Pyrenees. *Climatic Change*, Vol.63, No.1-2, pp. 181-200, ISSN 1573-1480

Čerňanský, J. & Kožuch, M. (2001). The monitoring of changes high mountains landscape national park nízke tatry with methods digital photogrammetry. *Geodetický a kartografický obzor*, Vol.47, No.89, pp. 242-249, ISSN 0016-7096

Daniels, L. D. & Veblen, T. T. (2004). Spatioteporal influences of climate on altitudinal treeline in Northen Patagonia. *Ecology*, Vol.85, No.5, pp. 1284–1296, ISSN 0012-9658

Diaz, H. F. & Bradley, R. S. (1997). Temperature variations during the last century at High elevation sites. *Climatic Change*, Vol.36, No.3-4, pp. 253–279, ISSN 1573-1480

Dirnböck, T., Dullinger, S. & Grabherr, (G. 2003). A regional impact assessment of climate and land-use change on alpine vegetation. *Journal of Biogeography*, Vol.30, No.3, pp. 401–417, ISSN 1365-2699

Dirnböck, T., Essl, F. & Rabitsch, W. (2011). Disproportional risk for habitat loss of high-altitude endemic species under climate change. *Global Change Biology*, Vol.17, No.2, pp. 990–996, ISSN 1365-2486

Dullinger, S., Dirnböck, T. & Grabherr, G. (2003a). Patterns of Shrub Invasion into High Mountain Grasslands of the Northern Calcareous Alps, Austria. *Arctic, Antarctic, and Alpine Research*, Vol.35, No.4, pp. 434-441, ISSN 1938-4246

Dullinger, S., Dirnböck, T., Greimler, J. & Grabherr, G. (2003b). A resampling approach for evaluating effects of pasture abandonment on subalpine plant species diversity. *Journal of Vegetation Science*, Vol.14, No.2, pp. 243–252, ISSN 1654-1103

Dullinger, S., Dirnböck, T. & Grabherr, G. (2004). Modelling climate change-driven treeline shifts: relative effects of temperature increase, dispersal and invasibility. *Journal of Ecology*, Vol.92, No.2, pp. 241–252, ISSN 1365-2745

Dulinger, S., Dirnböck, T., Köck, R., Hochbichler, E., Englisch, T., Sauberer, N. & Grabherr, G. (2005). Interactions among tree-line conifers: differential effects of pine on spruce and larch. *Journal of Ecology*, Vol.93, No.5, pp. 948–957, ISSN 1365-2745

Falťan, V. & Saksa, M. (2007). Zmeny krajinnej pokrývky okolia Štrbské pleso po veternej kalamite v novembri 2004. In: *Geografický časopis*, Vol.59, No.4, pp. 359-372, ISSN 0016-7193

Faško, P., Lapin, M. & Pecho, J. (2008). 20-year extraordinary climatic period in Slovakia. *Meteorologický časopis*, Vol.11, No.3, pp. 99–105, ISSN 1335-339X

Feranec, J. Cebecauerova, M., Cebecauer, T., Husár, K., Oťaheľ, J., Pravda, J. & Súri, M. (1997). Analýza zmien krajiny aplikáciou údajov diaľkového prieskumu zeme. *Geographia Slovaca*, Vol.13, pp. 1-64, ISSN 1210-3519

Gehrig-Fasel, J., Guisan, A. & Zimmermann, N. E. (2007). Tree line shifts in the Swiss Alps: Climate change or land abandonment? *Journal of Vegetation Science*, Vol.18, No.4, pp. 571–582, ISSN 1654-1103

Giorgi, F., Hurrell, J. W., Marinucci, M. R. & Beniston, M. (1997). Elevation dependency of the surface climate change signal: a model study. *Journal of Climate*, Vol.10, No.2, pp. 288–296. ISSN 1520-0442.

Gottfried, M., Pauli, H., Reiter, K. & Grabherr, G. (1999). A fine-scaled predictive model for changes in species distribution patterns of high mountain plants induced by climate warming. *Diversity and Distributions*, Vol.5, No.6, pp. 241–251, ISSN 1472-4642

Grabherr, G., Gottfried, M., Gruber, A. & Pauli, H. (1995). Patterns and Current Changes in Alpine Plant Diversity. In: *Arctic and alpine biodiversity: patterns, causes and ecosystem consequences, Ecological Studies, Vol.113*, Chapin III, F. S, Koerner C. (Ed.), pp. 167-181, Springer, ISBN 3540579486, Berlin, Heidelberg, Germany

Harsch, M. A., Hulme, P. E., McGlone, M. S. & Duncan, R. P. (2009). Are treelines advancing? A global meta-analysis of treeline response to climate warming. *Ecology Letters*, Vol.12, No.10, pp. 1040–1049, ISNN 1461-0248.

Heywood, D. I., Price, M. F. & Petch J. R. (1994). Mountain regions and geographic information systems: an overview (Ed.), In: *Mountain Environments and Geographic Information Systems*, Price, M. F. (Ed.), pp. 1-23, Taylor & Francis, ISBN 07484-0088-5, London, United Kingdom

Hreško, J. & Boltižiar, M. (2001). The influence of the morphodynamic processes to landscape structure in the high mountains (Tatra Mts.). *Ekológia*, Vol.20, No.3, pp. 141-149, ISSN 1377-947X

Huntley, B. & Baxter, R. (2002). Climate change and wildlife conservation in the British uplands. In: *The British Uplands: Dynamics of Change Report Number 319*, Burt, T.P., Thompson, D.B.A. & Warburton, J. (Ed.), pp. 41-47, Joint Nature Conservation Committee, ISSN 0963-8091, Peterborough

Chomitz, K. & Šamaj, F. (1974). Precipitation characteristics. In: *Climate of Tatras*, Konček, M Bohuš, I., Briedoň, V., Chomicz, K., Intribus, R., Kňazovický, L., Kolodziejek, M., Kurpelová, M., Murinová, G., Myczkowski, Š., Orlicz, M., Orliczowa, J., Otruba, J., Pacl, J., Peterka, V., Petrovič, Š., Plesník, P., Pulina, M., Smolen, F., Sokolowska, J., Šamaj, F., Tomlain ,J., Volfová, E., Wiszniewski, W., Wit-Józwikowa, K., Zych, S. & Žák, B. (Ed.), pp 443–536, Veda, Bratislava, Slovakia

IPCC 2007. Climate Change (2007). The Physical Science Basis. Contribution of Working Group I to the Fourth Assessment Report of the Intergovernmental Panel on Climate Change, Solomon, S., Qin, D., Manning, M., Chen, Z., Marquis, M., Averyt, K.B., Tignor M. & Miller, H.L., pp. 996, ISBN 978 0521 88009-1, Cambridge University Press, Cambridge, United Kingdom and New York, NY, USA,

Jodłowski, M. (2006). Anthropogenic transformation of the krummholz-line in some mountain ranges in Central Europe. In: *Global Change in Mountain Regions*, Price, M. (Ed.), pp 186–187, Sapiens Publishing, London, United Kingdom

Jodłowski, M. (2007). Krummholz-line in the Tatra Mts. Babia Góra and the Giant Mts. Ecotone Structure and Dynamics [in Polish] [PhD dissertation]., Poland Institute of Geography and Spatial Management of the Jagiellonian University, Krakow, Poland

Kohút, F. (2006). Vzťah krajinnej pokrývky k vybraným charakteristikám reliéfu v Západných Tatrách (Jalovecká Dolina). In: *Aktivity v kartografii 2006*, Feranec, J. & Pravda, J. (Ed.), pp. 129-136, Kartografická spoločnosť slovenskej republiky a Geografický ústav SAV, ISBN 80-89060-09-94, Bratislava, Slovakia

Konček, M. & Orlicz, M. (1974). Temperature characteristics. In: *Climate of Tatras*, Konček M, Bohuš I., Briedoň V., Chomicz K., Intribus R., Kňazovický L., Kolodziejek M., Kurpelová M., Murinová G., Myczkowski Š., Orlicz M., Orliczowa J., Otruba J., Pacl J., Peterka V., Petrovič Š., Plesník P., Pulina M., Smolen F., Sokolowska J., Šamaj F., Tomlain J., Volfová E., Wiszniewski W., Wit-Józwikowa K., Zych S. & Žák B. (Ed.), pp. 89–179, Veda, Bratislava, Slovakia

Körner, C. & Paulsen, J. (2004). A world-wide study of high altitude treeline temperatures. *Journal of Biogeography*, Vol.31, No.5, pp. 713–732, ISSN 1365-2699

Kullman, L. (2002). Rapid recent range-margin rise of tree and shrub species in the Swedish Scandes. *Journal of Ecology*, Vol.90, No.1, pp. 68–77, ISSN 1365-2745

Lajczak, A. (1996). Hydrology, In: *Nature of the Tatra National Park*, Mirek, Z., Głowaciński, Z., Klimek, K., Piękoś-Mirkowa, H., pp 169–196, Tatrzański Park Narodowy, Zakopane-Krakow, Poland

Lapin, M., Šťastný, P. & Chmelík, M. (2005). Detection of climate change in the Slovak mountains. *Croatian Meteorological Journal*, Vol.40, No.40, pp. 101–104, ISSN 1330-0083

Lapin, M., Faško, P. & Peclio, J. (2007). Snow cover variability and trends in the Tatra Mountains in 1921-2006, *Proceedings of the 29th International Conference on Alpine Meteorology Chambéry*, France, 4.-8. June 2007

Lapin, M., Gera, M., Hrvoľ, J., Melo, M. & Tomlain, J. (2009). Possible impacts of climate change on hydrologic cycle in Slovakia and results of observations in 1951–2007. *Biologia*, Vol.64, No.3, pp. 454-459, ISSN 1335-6399

Martazinova, V., Ivanova, O. & Shandra, O. (2009). Climate and treeline dynamics in the Ukrainian Carpathians, In: *Sustainable Development and Bioclimate Reviewed Conference, Proceeding 5th to 8th October 2009*, Pribullová, A. & Bičárová, S., (Ed.), pp. 105-106, Geophysical Institute of the SAS, CD ISBN 978-80-900450-1-9, Stará Lesná, Slovakia, 5-8. October 2009

Melo, M. (2005). Warmer periods in the Slovak mountains according to analogue method and coupled GCM. *Croatian Meteorological Journal*, Vol.40, No.40, pp. 589-592, ISSN 1330-0083

Melo, M. (2007). Regional climatic scenarios for Slovak mountain region based on three global GCMs. *29th International Conference on Alpine Meteorology*, Chambéry, France, June, 2007

Melo, M., Lapin, M. & Damborska, I. (2009). Methods for the design of climate change scenario in Slovakia for the 21st century. *Bulletin of Geography – physical geography series*, Vol.1, No.1, pp. 77–90, ISSN 2080-7686

Menzel, A. & Fabian, P. (1999). Growing season extended in Europe. *Nature*, Vol.397, No.6721, pp. 397-659, ISSN 0028-0836

Mihai, B, Savulescu, I. & Sandric, I. (2007). Change detection analysis (1986–2002) of vegetation cover in Romania. *Mountain Research and Development*, Vol.27, No.3, pp. 250–258, ISSN 1994-7151

Minďáš, J., Lapin, M. & Škvarenina, J. (1996). *Klimatické zmeny a lesy Slovenska*. In: *Národný klimatický program Slovenskej republiky*, MŽP SR, Bratislava, Slovakia

Minďáš, J., Škvarenina, J., Ďurský, J., Lapin, M., Marečková, K., Priwitzer, T., Střelcová, K., Šály, R., Pavlenda, P., Tomlain, J., Čaboun, V., Jankovič, J., Merganič, J., Miková, A., Novotný, J., Jakuš, R., Kmeť, J., Midriak, R., Vladovič, J., Konôpka, J., Kunca, A., Leontovyč, R., Turčáni, M., Zúbrik, M. & Finďo, S. (2003). *Forests of Slovakia and global climatic changes*. EFRA Zvolen & LVÚ Zvolen, ISBN 80-228-1209-9, Zvolen, Slovakia

Minďáš, J., Čaboun, V. & Priwitzer, T. (2004). Horná hranica lesa a očakávané klimatické zmeny. In *Zborník Turiec a Fatra*, Kadlečík, J. (Ed.), pp. 17-23, ŠOP SR, ISBN 8089035302, Vrútky, Slovakia

Moen, J. (2006). Treeline dynamics in a changing climate. In: *Global Change in Mountain Regions*, Price, M. (Ed.), pp. 187–188. Sapiens Publishing, London, United Kingdom

Nemčok, J., Bezák, V., Janák, M., Kahan, Š., Ryja, W., Kohút, M., Lehotský, I., Wieczorek, J., Zelman, J., Mello, J., Halouzka, R., Raczkowski, W. & Reichwalder, P. (1993). *Vysvetlivky ku geologickej mape Tatier*, Geologický ústav Dionýza Štúra, ISBN 8085314231, Bratislava, Slovakia

Nicolussi, K. & Patzelt, G. (2006). Klimawandel und Vera¨nderungen an der alpinen Waldgrenze-Aktuelle Entwicklungen im Vergleich zur Nacheiszeit. *BFW-Praxisinformation* Vol.10, No.1, pp. 3-5, ISSN 1815-3895

Olah, B., Boltižiar, M., Petrovič, F. & Gallay, I., (2006). *Vývoj využitia krajiny slovenských biosférických rezervácií UNESCO*, B. TU a SNK MaB, ISBN 80-228-1695-7, Zvolen, Slovakia

Olah, B. & Boltižiar, M., (2009). Land use changes within the Slovak biosphere reserves' zones. *Ekológia*, Vol.28, No.2, pp. 127-142 ISSN 1335-342X

Ostrožlík, M. (2008). Atmospheric precipitation at Skalnaté pleso and Lomnický štít, In: *Bioklimatologické aspekty hodnocení procesů v krajině*, ISBN 978-80-86690-55-1, Mikulov, Czech Republic, 9. – 11. September 2008

Ostrožlík, M. (2010). Variability of the air temperature and atmospheric precipitation in the high-mountain positions of the Low and High Tatras in winter. *Contributions to Geophysics and Geodesy*, Vol.40, No.1, pp. 87–101, ISSN 1338-0540

Parmesan, C. & Yohe, G. (2003). A globally coherent fingerprint of climate change impacts across natural systems. *Nature*, Vol.421, No.6918, pp. 37–42, ISSN 0028-0836

Pauli, H., Gottfried, M. & Grabherr, G. (1996). Effects of climate change on mountain ecosystems - upward shifting of alpine plants. *World Resource Review*, Vol.8 No.3, pp. 382- 390, ISSN 1042-8011

Peñuelas, J., Ogaya, R., Boada, M. & Jump, S. A. (2007). Migration, invasion and decline: changes in recruitment and forest structure in a warming-linked shift of European beech forest in Catalonia (NE Spain). *Ecography*, Vol.30, No.6, pp. 829–837, ISSN 1600-0587

Root, T., Price, J. T., Hall, K. R. Schneider, S. H., Rosenzweig, C. & Pounds, J. A., (2003). Fingerprints of global warming on wild animals and plants. *Nature*, Vol.421, No.6918, pp. 57–60, ISSN 0028-0836

Rossi, S., Deslauriers, A., Anfodillo, T. & Carraro, V. (2007). Evidence of threshold temperatures for xylogenesis in conifers at high altitudes. *Oecologia*, Vol.152, No.1, pp. 1–12, ISSN 1432-1939

Sætersdal, M., Birks, H. J. B., & Peglar, S. M. (1998). Predicting changes in Fennoscandian vascular-plant species richness as a result of future climatic change. *Journal of Biogeography*, Vol.25, pp. 111–122, ISSN 1365-2699

Sitko, A. & Troll, M. (2008). Timberline changes in relation to summer farming in the western Chornohora (Ukrainian Carpathians). *Mountain Research and Development*, Vol.28, No.3-4, pp. 263–271, ISSN 1994-7151

Smith, W. K., Germino M. J., Hancock, T. E. & Johnson, D. M. (2003). Another perspective on altitudinal limits of alpine timberlines. *Tree Physiogy*, Vol.23, No.16, pp. 1101-1112, ISSN 1758-4469

Smith, W. K., Germino M. J., Hancock, T. E., Johnson, D. M. & Reinhardt, K. (2009). The Altitude of Alpine Treeline: A Bellwether of Climate Change Effects. *The Botanical Review*, Vol.75, No.2, pp. 163–190, ISSN 1874-9372

Song, M., Zhou, C. & Ouyang, H. (2004). Distributions of dominant tree species on the Tibetan Plateau under current and future climate scenarios. *Mountain Research and Development*, Vol.24, No.2, pp. 166–173, ISSN 1994-7151

Škvareninová, J. (2009). Priebeh vegetatívnych fenologických fáz autochtónnych populácií smreka obyčajného (*Picea abies* [L.] Karst.) *Lesnícky časopis - Forestry Journal*, Vol.55, No.1, pp. 13 -27, ISSN 0323-104

Škvareninová, J. (2008). Start of spring phenophases in pedunculate oak (*Quercus robur* L.) in the Zvolenska basin, in relation to temperature sums. *Meteorological Journal*, Vol.11, No.1–2, pp. 15–20, ISSN 1335-339X

Švajda, J., Solár, J., Janiga, M. & Buliak M. (2011). Dwarf Pine (*Pinus mugo*) and Selected Abiotic Habitat Conditions in the Western Tatra Mountains. *Mountain Research and Development*, Vol.31, No.3, pp. 220-228, ISSN 1994-7151

Theurillat, J-P. & Guisan, A. (2001). Potential Impact of Climate Change on Vegetation in the European Alps: A Review. *Climatic Change*, Vol.50, No.1-2, pp. 77–109, ISSN 1573-1480

Theurillat, J-P., Felber, F., Geissler, P., Gobat, J-M., Fierz, M., Fischlin, A., Küpfer, P., Schlüssel, A., Velluti, C., Zhao, G-F. & Williams, J. (1998). Sensitivity of plant and soil ecosystems of the Alps to climate change, In: *Views from the Alps: Regional Perspectives on Climate Change*. Cebon, P., Dahinden, U., Davies, H.C., Imboden, D. & Jaeger, C.C. (Ed.), pp. 225-308, MIT Press, ISBN 9780262032520, Cambridge, United Kingdom

Visser, M. E. & Both, Ch. (2005). Shifts in phenology due to global climate change: the need for a yardstick. *Proceeding of Royal Society B*, Vol. 272, No.1581, pp. 2561–2569, ISSN, 1471-2954

Vittoz, P., Rulence, B., Largey, T. & Freléchoux, F. (2008). Effects of Climate and Land-Use Change on the Establishment and Growth of Cembran Pine (*Pinus cembra* L.) over the Altitudinal Treeline Ecotone in the Central Swiss Alps. *Arctic, Antarctic, and Alpine Research*, Vol.40, No.1, pp. 225-232, ISSN 1938-4246

Vojtek, M., Faško, P. & Šťastný, P. (2003). Some selected snow climate trends in Slovakia with respect to altitude. *Acta Meteorologica Universitatis Comenianae*, Vol.32, No.1, pp. 17 – 27, ISSN 0231-8881

Vološčuk, I, Bohuš, I., Bublinec, E., Bohušová-Hradiská, H., Drdoš, J., Dubravcová, Z., Gáer J., Greguš, C., Haková, J., Chovancová, B., Hindák, F., Janiga, M., Kocián, Ľ., Korbel, L., Koreň, M., Kováč, J., Kyselová, Z., Lazebníček, J., Linkeš, V., Ložek, V., Majzlan, O., Marenčák, M., Midriak, R., Nemčok, J., Novák, V., Olejník, J., Ostrožlík, M., Pacl, J., Paclová, L., Schmidt, M., Smolen, F., Spitzkopf, P., Šoltésová, A., Šoltés, R. & Šomšák, L. (1994). *Tatra National Park – Biosphere Reserve*, Gradus, ISBN 8090139248, Martin, Slovakia

An Illustration of the Effect of Climate Change on the Ocean Wave Climate - A Stochastic Model

Erik Vanem, Bent Natvig, Arne Bang Huseby and Elzbieta M. Bitner-Gregersen

Additional information is available at the end of the chapter

1. Introduction

Rough seas are a major cause for ship losses and significantly contribute to the risk to maritime transportation. It is therefore important to take severe sea state conditions into account, with due treatment of the uncertainties involved, in ship design and operation. There is thus a need for appropriate stochastic models describing the variability of sea states and these should also consider long-term trends related to climate change. This chapter presents such a stochastic model, aiming at describing the spatial and temporal variability, as well as long-term trends, in the ocean wave climate.

The stochastic ocean wave model presented in this chapter exploits the flexible framework of Bayesian hierarchical space-time models. It allows modelling of complex dependence structures in space and time and incorporation of physical features and prior knowledge, yet at the same time remains intuitive and easily interpreted. Furthermore, by taking a Bayesian approach, the uncertainties of the model parameters are also taken into account. Different alternatives for modelling the long-term trend are suggested, with and without a regression component with CO_2 as an explanatory variable. The models have been fitted by monthly maximum significant wave height data for an area in the North Atlantic ocean. The different components of the model will be outlined in this chapter, and the results will be discussed. Furthermore, the influence of the estimated expected long-term trends on the environmental loads of ocean-going ships will be investigated.

According to the Intergovernmental Panel of Climate Change (IPCC) [21–23], the globe is experiencing climate change. The IPCC report [21] also presents projections of future climate change, and it is deemed very likely that frequencies and intensities of some extreme weather events will increase. However, a more recent summary report is more careful in its conclusions [23].

Ships and other marine structures are constantly exposed to the wave and wind forces of its environment, and extreme ocean climate represents a risk to marine operations. Bad weather is indeed often involved in accidents and ship losses, and this stresses the importance of taking extreme sea state conditions adequately into account in ship design. This is important to ensure that the ships can withstand the environmental forces they are expected to encounter throughout their lifetime. Hence, a correct and thorough understanding of meteorological and oceanographic conditions and the extreme values of relevant wave and wind parameters, in particular wave parameters such as the significant wave height (H_s) is of paramount importance to maritime safety, and there is a need for appropriate statistical models to describe the variability of these phenomena. Long-term statistics can then be combined with individual wave statistics in order to estimate the highest waves that should be used in design of marine structures, as outlined in e.g. [1].

In particular, with the observed and projected climate change that the globe is currently experiencing, it may no longer be sufficient to base design codes and safety standards on current knowledge about the past and present ocean environment. The implicit assumption that the future will be like the past may no longer be even approximately valid and there is a need to consider how wave parameters are expected to change in the future, as a consequence of climate change. Thus, there is a need for time-dependent statistical models that can take the long-term time-dependency of integrated wave parameters properly into account.

In this chapter, a Bayesian hierarchical space-time model ([46], [47]) will be outlined that has been developed to describe significant wave height as a stochastic spatio-temporal process ([37]). The model is hierarchical and allows for modelling of complex dependence structures in space and time and includes prior information by way of informative priors. It is built up by different components including a purely spatial field, a short-term, spatio-temporal dynamic component, a temporal seasonal component, and finally, a separate term for modelling long-term trends, possibly as a consequence of climate change. The model has been fitted to significant wave height data for an area in the North Atlantic ocean, selected because North Atlantic conditions are used as design basis for the majority of sailing ships. The selection of the modelling approach was based on a thorough literature survey, presented in [35]. Bayesian hierarchical space-time models are also treated in the book [13].

The model and its various components will be outlined in a subsequent section. Furthermore, different variations and extensions to the main model will be introduced. Most importantly, a logarithmic transform of the data yields a different interpretation of the model ([39]) and the long-term trend component will be modelled as a regression block, where the trends in significant wave height are regressed on levels of atmospheric CO_2 ([38]). In this way, long-term trends in the data are identified, and projections of future ocean wave climate can be made based on different emission scenarios.

Finally, it is demonstrated how the estimated expected increase in severity of future ocean wave climate is related to the structural loads and responses of ships at sea and how these effects can be taken into account in load calculations ([36]). It was found that the models predict a non-negligible effect on the extreme environmental loads. Hence, the findings indicate that the effect of climate change on the ocean wave climate may need to be considered in ship and marine structures design.

2. Description of location

The scope of this study is restricted to consider an area in the North Atlantic ocean, i.e. the ocean area between 51°-63° north and 12°-36° west (or 324°-348° east). The spatial resolution of the data is $1.5° \times 1.5°$, hence a grid of $9 \times 17 = 153$ datapoints covers the area. It is noted that due to the curvature of the surface of the earth, the distance between gridpoints will not be constant throughout the area. The distance in the north-south direction is fairly constant but the distance in the longitudinal direction (east-west) differs significantly for different latitudes. However, in the following analysis of spatial variability, this fact will be ignored. The area under consideration is illustrated on a map in figure 1.

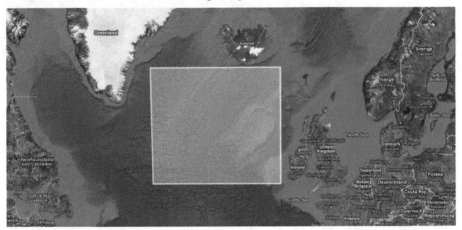

Figure 1. The area of the North Atlantic ocean under consideration

3. Description of data

Data for significant wave height have been used to fit the stochastic model, and data on levels of CO_2 concentrations in the atmosphere have been used as covariates. In the following, a brief description of these sets of data will be given.

3.1. Wave data

The reanalysis project ERA-40 [34] was carried out by the European Centre for Medium-Range Weather Forecasts (ECMWF) and covers the 45-year period from September 1957 to August 2002. Data obtained from this reanalysis include six-hourly sampled global fields of significant wave height; global, continuous data are available on a $1.5° \times 1.5°$ grid, making this perhaps the most complete wave dataset available to date.

It has been reported that the ERA-40 dataset contains some limitations which indicate problems in using these data for modelling long-term trends in extreme waves ([32]). However, corrected datasets for the significant wave height have been produced, resulting in a new 45-year global six-hourly dataset of significant wave height ([10]). When compared to buoy measurement and global altimeter data, this corrected dataset, referred to as the C-ERA-40 data, shows clear improvements compared to the original data ([11]). It is this corrected dataset, which was kindly provided by the Royal Netherlands Meteorological Institute (KNMI)[1] that has been used in this study. It includes fields of significant wave height sampled every 6th hour with a spatial resolution of $1.5° \times 1.5°$ covering the period from January 1958 to February 2002 (i.e. a total of 44 years and 2 months which corresponds to a sequence of 64 520 points in time). However, for this particular study it is deemed sufficient to use monthly maximum data at each location, totalling 530 monthly maxima in time for each location.

In general, it is acknowledged that wave buoys are regarded as highly accurate instruments, and it is stated in e.g. [7] that both the systematic and random error of significant wave height measurements by buoys are negligible. However, when calibrating hindcast data against observations, the data will still be subject to epistemic uncertainty due to the way the calibration is carried out and high values of significant wave height will normally be more affected by uncertainties, as discussed in [6]. For the purpose of this study it is emphasized that all modelling and all results are conditional on the input data and data validation and data uncertainty is considered out of scope.

3.2. CO_2 data

Concentrations of atmospheric CO_2 have been used as covariates for explaining possible long-term trends in the significant wave height, and basically two sets of data have been exploited; historic data for model fitting and projections of future concentration levels for future predictions.

3.2.1. Historic CO_2 data

The aim of introducing a regression component with CO_2 levels as covariates is to identify long-term trends, and it is deemed sufficient to use monthly data. Hence, monthly average CO_2 data from the Mauna Loa Observatory, Hawaii, have been used ([33]). The data are on the format of the number of molecules of carbon dioxide divided by the number of molecules of dry air multiplied by one million (parts per million = ppm), and data are available from March 1958 to present. The data set contains the monthly averages determined from daily averages, as well as interpolated monthly averages where missing data have been replaced by interpolated values. Finally, monthly trend values are given where the seasonal cycle has been removed and where linear interpolation has been used for missing months. For the purpose of this study, the monthly trend time series will be used as covariates for the long term trend. The seasonal cycle in the monthly maximum significant wave height is accounted for in a separate seasonal component in the model.

The monthly interpolated and trend data are illustrated in the graphs in figure 2 and the vertical lines represent the part of the time series that overlap the C-ERA-40 data for significant wave height. It is noted that the CO_2 data starts at March 1958 whereas the significant wave

---

[1] Private communication with Dr. Andreas Sterl, KNMI

height data starts at January 1958. Therefore, when utilizing the CO_2 data, the model will be run with data starting at January 1959.

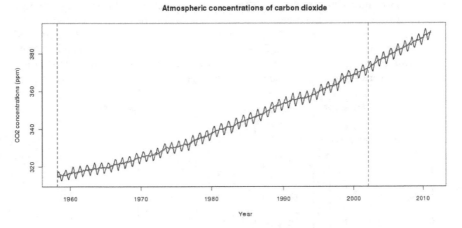

Figure 2. CO_2 data from the Mauna Loa Observatory, monthly interpolated data (black line) and trend data with seasonal effects removed (red lines)

It is acknowledged that CO_2 is just one greenhouse gas (GHG) and that it does not alone determine the radiative forcing of the globe; other important GHGs are for example methane (CH_4) and nitrous oxide (N_2O). Nevertheless, it is generally agreed that CO_2 is the most important GHG and for the purpose of this study, it is construed as a proxy for the concentration of GHG in the atmosphere. More sophisticated models could include other GHGs and aerosols as covariates as well. It is also noted that the data stem from observations outside of the area in the North Atlantic which is the focus of this study. However, it is assumed that CO_2 is well mixed in the atmosphere, and that this does not introduce any notable bias in the results pertaining to expected long-term trends.

3.2.2. Future projections of CO_2 levels

Future projections of the atmospheric concentration of CO_2 will be exploited to make projections of future wave climate. Future predictions are inevitably uncertain, and different projections of CO_2 levels have been made based on different emission scenarios ([27]). Projected emissions and concentrations presented by IPCC for the four marker scenarios A1B, A2, B1 and B2 have been considered[2]. The scenarios A2 and B1 correspond to the highest and lowest projected CO_2 levels respectively, and it is therefore assumed sufficient to employ these two in the modelling. Scenario A2 might be an extreme scenario, but from a precautionary perspective it is important to concider since this could be construed as a worst case scenario. The CO_2 projections data can also be found in appendix II of [20].

---

[2] The IPCC Data Distribution Centre, URL: http://www.ipcc-data.org/ddc_co2.html

The projected levels of atmospheric CO_2 concentrations are given for every ten year towards 2100. For the purpose of this study, monthly averages are needed, and simple linear interpolation within each decade has been used to estimate monthly projections. The decadal projections are then assumed as the value for January of that year. In this way, monthly projections of CO_2 levels in the atmosphere from year 2010 until 2100 is obtained for use as covariates. For the years 2002 to 2010, where actual observations are available, recorded monthly averages from the Mauna Loa Observatory will be used. The interpolated monthly projections are plotted together with the original decadal projections in figure 3 (the vertical bars in the plots correspond to the decadal reference projections from the ISAM model ([24])).

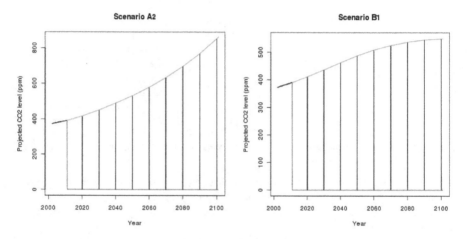

Figure 3. Interpolated monthly CO_2 level projections for scenarios A2 and B1

The uncertainty of the data is not accounted for and any results are also conditional on the data used for the covariates. Uncertainties are of course large for future predictions, but it is assumed that the projections suggested by the IPCC correspond to the best current knowledge available. The uncertainties of future projections of CO_2 concentrations were discussed in e.g. [25] and it was suggested to assign probabilities for the various scenarios. However, such probabilities have not been assigned in this study. The historic data and the projections corresponding to the four marker scenarios are illustrated together in figure 4.

4. Initial inspection of the wave data

The density of all the monthly maxima is shown in figure 5 and two distinct modes can be identified, one around 5 meters and another at about 8 meters. It is believed that these correspond to different characteristics during calm and rough seasons. For the whole dataset, the mean monthly maximum is 7.5 meters, and the average monthly maxima for each month are given in table 1. Density plots (not shown) for each month show that the months January - March and October-December have peaks around 8-9 meters and that the months May-August have peaks around 5 meters. The remaining months, April and September are

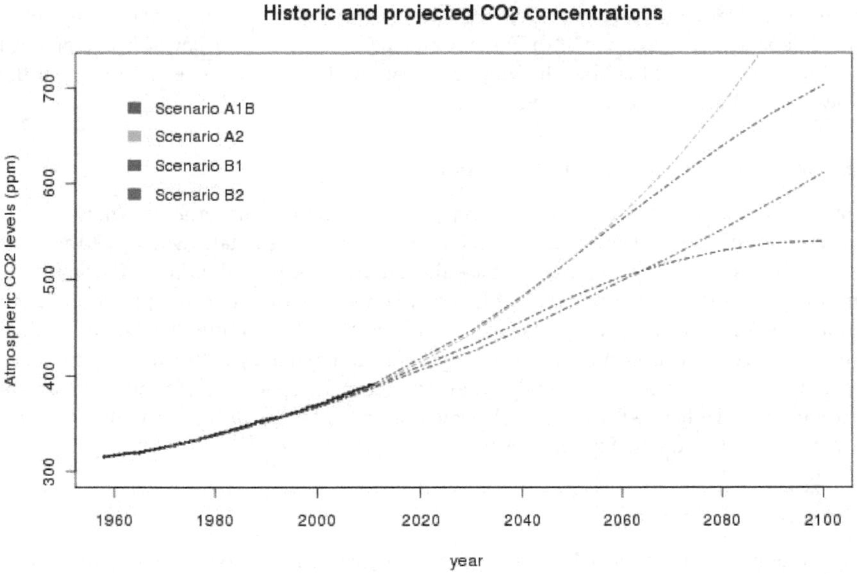

Historic and projected CO2 concentrations

Figure 4. Atmospheric CO_2 levels: Historic data and future projections

more flat with most probability mass between 5 and 8 meters. At any rate, the two modes in the density plot seem to be explained by the peaks at the different months.

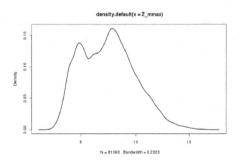

density.default(x = Z_mmax)

N = 61090   Bandwidth = 0.2323

Figure 5. The density of the monthly maximum data

Jan	Feb	Mar	Apr	May	Jun	Jul	Aug	Sep	Oct	Nov	Dec
9.87	9.63	8.91	7.18	5.89	5.03	4.42	5.04	6.96	8.21	8.69	9.79

Table 1. Average monthly maxima for each month

One may also check for normality or log-normality, but tests show that the data are neither Gaussian nor log-normal. Furthermore, attempts to describe the spatial and temporal variability by simple regression and autocorrelation models fail. Hence, it is apparent that the data cannot be well described by simple models and a somewhat more sophisticated model must be constructed. Hierarchical models are known to model spatio-temporal processes with complex dependence structures at different scales [45]. Therefore, a Bayesian hierarchical space-time model, along the lines drawn out by e.g. [46] will be constructed to model the significant waveheight data in space and time.

4.1. Logarithmic transformation of the data

Higher wave heights are normally associated with higher uncertainty (noise), and heteroscedastic features are observed in the significant wave height data. One way to account for such heteroscedasticity could be to take the log-transform of the data. Furthermore, taking the log-transform of the data yields a fundamentally different interpretation of the contributions from the various model components, which become multiplicative rather than additive. Hence, a revised model would associate higher trends with extreme sea states compared to non-extremes. It is noted that for inference made on log-transformed data, biases may be introduced when re-transforming back to the original scale. Bias correction factors and how to deal with the re-transformation bias are discussed in [39].

5. The stochastic model

The spatio-temporal data will be indexed by two indices; an index x to denote spatial location with $x = 1, 2, \ldots, X = 153$ and an index t to denote a point in time with $t = 1, 2, \ldots, T = 530$ for monthly maximum data. The structure of the basic model, as well as a revised model for log-transformed data and an extended model with a regression block, will be outlined below.

5.1. Basic model

The basic model is similar to the model presented in [37], inspired by [28], and contains an observation model and several state models, as outlined below. All the stochastic terms introduced in the model are assumed mutually independent and independent in space and time, having a zero-mean Gaussian distribution with some random, but identical variance, i.e., with generic notation, $\varepsilon_\beta(x,t) \overset{i.i.d}{\sim} N(0, \sigma_\beta^2)$. It should be understood that the model is defined $\forall\, x \geq 1, t \geq 1$, as relevant for each component.

At the first level, the observations (monthly maximum significant wave height), Z at location x and time t, are modelled in the observation model as the latent variable, H, corresponding to the underlying significant wave height process, and some random noise, ε_z, which may be construed as statistical measurement error:

$$Z(x,t) = H(x,t) + \varepsilon_z(x,t) \tag{1}$$

The underlying process for the significant waveheight at location x and time t is modelled by the following state model, which is assumed split into a time-independent component, $\mu(x)$, a time- and space-dependent component $\theta(x,t)$ and spatially independent seasonal, $M(t)$, and long-term trend, $T(t)$, components as shown in eq. 2. The long-term temporal trend is assumed spatially invariant and is, in fact, the component of most interest as it models the effect of climate change on the ocean wave climate.

$$H(x,t) = \mu(x) + \theta(x,t) + M(t) + T(t) \tag{2}$$

The time-independent part is modelled as a first-order Markov Random Field (MRF), conditional on its nearest neighbours in all cardinal directions, and with different dependence parameters in lateral and longitudinal direction, as shown in eq. 3, with x^D = the location of the nearest gridpoint in direction D from x, where $D \in \{N, S, W, E\}$ and N = North, S = South, W = West and E = East. If x is at the border of the area, the value at the corresponding neighboring gridpoint outside the data area is taken to be zero.

$$\mu(x) = \mu_0(x) + a_\phi \left\{ \mu(x^N) - \mu_0(x^N) + \mu(x^S) - \mu_0(x^S) \right\}$$
$$+ a_\lambda \left\{ \mu(x^E) - \mu_0(x^E) + \mu(x^W) - \mu_0(x^W) \right\} + \varepsilon_\mu(x) \tag{3}$$

a_ϕ and a_λ are spatial dependence parameters in lateral (i.e. north-south) and longitudinal (i.e. east-west) direction respectively. The spatially specific mean, $\mu_0(x)$, is modelled as having a quadratic form with an interaction term in latitude and longitude. Letting $m(x)$ and $n(x)$ denote the longitude and latitude of location x respectively, it is assumed that

$$\mu_0(x) = \mu_{0,1} + \mu_{0,2}m(x) + \mu_{0,3}n(x) + \mu_{0,4}m(x)^2 + \mu_{0,5}n(x)^2 + \mu_{0,6}m(x)n(x) \tag{4}$$

The spatio-temporal dynamic term $\theta(x,t)$ is modelled as a vector autoregressive model of order one, conditionally specified on its nearest neighbours in all cardinal directions, as shown in eq. 5.

$$\theta(x,t) = b_0\theta(x,t-1) + b_N\theta(x^N,t-1) + b_E\theta(x^E,t-1)$$
$$+ b_S\theta(x^S,t-1) + b_W\theta(x^W,t-1) + \varepsilon_\theta(x,t) \tag{5}$$

b_0 as well as the parameters corresponding to the nearest neighbours, b_N, b_E, b_S, b_W are assumed invariant in space and are assumed to have interpretations connected to the underlying sea state dynamics.

The temporal component is modelled with a seasonal and a long-term trend part. The seasonal part is modelled as an annual cyclic contribution independent of space, see eq. 6. It has also been tried to include the second harmonic to account for semi-annual seasonal contributions, but these were found to be small compared to the annual contribution, as explained in [39].

$$M(t) = c\cos(\omega t) + d\sin(\omega t) + \varepsilon_m(t) \tag{6}$$

The long-term trend of the basic model is modelled as a simple Gaussian process with a quadratic trend, as shown in eq. 7. In [37], various model alternatives were suggested for this component, i.e. with linear and quadratic trends, but in this chapter, only the results pertaining to the linear models will be considered (model alternative 2 in [37]).

$$T(t) = \gamma t + \eta t^2 + \varepsilon_T(t) \tag{7}$$

5.2. Revised model for log-transformed data

With the log-transformed data, denoting $Z(x,t)$ the significant wave height at location x and time t, the log-transforms are first carried out for each location and time-point ([39]),

$$Y(x,t) = \ln Z(x,t) \tag{8}$$

Then, at the observation level, the log-transformed data, Y, are modelled as the latent (or hidden) variables, H, corresponding to some underlying significant wave height process, and some random noise, ε_Y:

$$Y(x,t) = H(x,t) + \varepsilon_Y(x,t) \tag{9}$$

An equivalent representation of the observation model would be

$$Z(x,t) = e^{H(x,t)}e^{\varepsilon_Y(x,t)}, \tag{10}$$

where now the noise term has become a multiplicative factor rather than an additive term and, conditioned on $H(x,t)$, the significant wave height $Z(x,t)$ will be log-normally distributed.

The underlying process for the significant wave height at location x and time t is modelled by the state model which is identical to the state model for the basic model in the preceding section, but it corresponds to the alternative representation in eq. 11 on the original scale; the significant wave height can be written as the product of five multiplicative factors and therefore, the contribution from each of the model components will have a fundamentally different interpretation compared to the model for the original data.

$$Z(x,t) = e^{\mu(x)}e^{\theta(x,t)}e^{M(t)}e^{T(t)}e^{\varepsilon_Y(x,t)} \tag{11}$$

In particular, the long-term trend will be modelled as a multiplicative factor, meaning that a higher trend will be ascribed to more severe sea states, i.e. extremes will be modelled with a higher trend than non-extremes. This feature was also reported by e.g. [49].

The same model alternatives as for the basic model have been tried out, but in this chapter, again only the results pertaining to the linear model will be reported.

5.3. Extended model with a regression component

Having established the basic model and found it to perform well for the significant wave height data, a model extention is introduced, where the long-term trend component $T(t)$ in eq. 7 is replaced by the regression component in eq. 12 ([38]).

$$T(t) = \gamma G(t) + \eta \ln G(t) + \delta G(t)^2 + \varepsilon_T(t) \tag{12}$$

With this model, the long-term trend in monthly maximum significant wave height is regressed on CO_2 concentrations in the atmosphere, assuming first a combination of a linear, square and logarithmic trend with respect to the level of CO_2. $G(t)$ denotes the average level of CO_2 in the atmosphere at time t. It is noted that CO_2 is known to mix well in the atmosphere, so there are no spatial description of this regression term. Different alternatives for the stochastic relationship were tried out in [38], but in this chapter only results pertaining to the model alternative performing best will be presented, i.e. the model with a combination of a linear and logarithmic relationship (setting $\delta = 0$).

5.4. Critical assumptions and prior distributions

The models presented in this chapter are stochastic models and as such they are simplified representations of the real world. All models imply simplifications and rely on a set of critical assumptions. The validity of those assumptions determines how well the model describes reality, and it is important to be aware of the most crucial model assumptions.

All the models presented in this chapter consist of different components in space and time, and an implicit assumption is that this separation of the significant wave height process into different contributions is reasonable. For example, this means that all long-term trends can be described by the separate long-term component. Also, the assumption of independent Gaussian noise associated with the various components is essential to the statistical model, but this assumption can be checked by way of normal probability plots of the residuals.

The extended model uses a regression component towards CO_2 to describe long-term variation in the ocean wave climate. Hence, a very critical model assumption is the stochastic dependence between levels of CO_2 in the atmosphere and the ocean wave climate. It is assumed that there is such a stochastic dependence and this might be a realistic assumption, as increased levels of CO_2 in the atmosphere are associated with higher temperatures, more energy in the weather systems and consequently rougher wave climate. However, it is further assumed that this stochastic dependence structure will remain essentially unchanged over time, from the past into the future. This is of course a critical assumption inherent in the model and any results are conditional on this assumption being realistic.

Furthermore, it is assumed that the CO_2 projections are reliable and results are conditional on the CO_2 data that has been utilized. In particular, no particular attention has been drawn towards possible climate tipping points or other effects that may skew the correlation between CO_2 levels in the atmosphere and ocean waves, and this introduces considerable uncertainty that has not been accounted for. Notwithstanding, the models presented herein are still believed to be interesting to investigate and they explore future ocean wave climate based on the best available knowledge of the future levels of CO_2 as a result of various emission scenarios.

Only CO_2 levels in the atmosphere have been considered, as a proxy of the level of greenhouse gases. It is normally considered that this is the dominant greenhouse gas, but omitting all other contributions is obviously a simplification. Furthermore, aerosols and other mitigating factors have not been considered as well as variability in solar radiation and external forcing.

It is noted that the model presented herein is a purely stochastic model, concerned with the stochastic dependencies in space and time, and the physics and regional characteristics of the wave climate are not modelled explicitly. However, it is argued that the physics underlying the wave generation process and all regional features are inevitably implicit in the data, and when applying the model on a particular data set any such physics and regional features would unavoidably be incorporated by way of the data. It is noted that the models can easily be updated to account for any known biases in the data.

Informative priors have been used in a Bayesian setting, where prior knowledge has been taken into account. Conditionally conjugate priors were adopted for each model parameter. For further details and exact values of the hyperparameters used for the priors, reference is made to [37–39]. However, it is argued that the results are not overly sensitive to the chosen prior distributions. It is well known in Bayesian analysis that the priors become asymptotically irrelevant as the amount of data increases, and the amount of data is quite large in this case.

5.5. Model comparison and prediction losses

Loss functions based on predictive power were constructed in order to compare model alternatives. Only one-step predictions are considered; the models are fitted with all data except for the last timepoint and predictions of the spatial field at this timepoint are compared to the data. The standard loss function in eq. 13 is defined where, for the timepoint selected for prediction, $Z(x)$ denotes the data at location x and $Z(x)_j^*$ denotes the predicted value of Z at location x in iteration j of the MCMC simulations.

$$L_s = \left[\frac{1}{Xn} \sum_{x=1}^{X} \sum_{j=1}^{n} \left(Z(x) - Z(x)_j^* \right)^2 \right]^{\frac{1}{2}} \tag{13}$$

One alternative loss function where the squared prediction errors have been weighted according to the actual observed significant wave height is also employed. A weight of size $Z(x)$ is included in order to give greater emphasis on prediction errors at locations where large significant wave heights are observed. Hence, an alternative loss function as given in eq. 14 is calculated.

$$L_a = \left[\frac{1}{n \sum_x Z(x)} \sum_{x=1}^{X} \sum_{j=1}^{n} Z(x) \left(Z(x) - Z(x)_j^* \right)^2 \right]^{\frac{1}{2}} \tag{14}$$

The predictions $Z(x)_j^*$ are taken as the estimated value of Z given the samples for all model parameters and variables in iteration j. The model specification gives

$$Z(x)_j^* = \mu(x)_j + \theta(x,t)_j + M(t)_j + T(t)_j + \varepsilon_Z(t)_j \tag{15}$$

for the basic and extended models and

$$Z(x)_j^* = e^{\mu(x)_j + \theta(x,t)_j + M(t)_j + T(t)_j + \varepsilon_Y(x,t)_j} \tag{16}$$

for the revised model for log-transformed data. The subscripts j denote the sampled parameters in iteration j. When using log-transformed data, the predictions are retransformed back to the original scale before the loss functions are calculated, i.e. the losses are on the same scale and should in principle be comparable although it is acknowledged that the comparison might not be completely fair for predictions made on a transformed scale.

6. Simulation results: trends and future projections

Posterior estimates of the parameters of the various models presented above have been obtained by Markov chain Monte Carlo methods, using the Gibbs sampler with additional Metropolis-Hastings steps (see e.g. [30]). Normal probability plots of the residuals indicate that the Gaussian model assumption is reasonable, and different informal tests suggest that the samples are from the stationary distribution, i.e. that the Markov chains have converged satisfactorily. Detailed results pertaining to all model components and posterior parameter estimates (mean and standard deviation), as well as descriptions of the MCMC settings, are presented in [37–40]. In this chapter, however, the main focus is on results pertaining to the long-term trends and expected future projections as a result of climate change.

6.1. Basic model

The basic model is found to perform reasonably well on the monthly maximum significant wave height data, with posterior estimates of the mean spatial field, $\mu(x)$, ranging from 6.1 to 7.3 meters over the area. The variability was greater in the north-south direction than in the east-west direction, which is reasonable. The expected contributions from the space-time dynamic part, $\theta(x, t)$ were between -1.1 and 1.8 meters and the expected seasonal contributions correspond to an annual cyclic variation of about $\pm$ 2.5 meters.

The component of most interest in this chapter, however, is the contribution from the long-term trend component $T(t)$, which is included to model any long-term effects, possibly related to global climate change. According to the linear model, an expected increase in monthly maximum significant wave height of 69 cm is estimated over the data-period. The 90% credible interval ranges from 45 - 94 cm, i.e. the complete interval is positive. These posterior trend contributions are illustrated in figure 6. The black line corresponds to the mean sampled $T(t)$, whereas the red lines correspond to the expected contribution γt as well as the 90% credible interval of the mean. The green line corresponds to no trend and it is clearly seen that the model detects a significant increasing trend in the wave climate.

In order to estimate future changes of the wave climate, possibly due to climate change, the estimated linear trend is extrapolated towards the year 2100. Hence, assuming that the identified long-term trend persists over 100 years, this would correspond to an expected increase in monthly maximum significant wave height of 1.6 meters over 100 years, with a 95% credibility of an increase of at least 1.0 meter.

6.2. Revised model - log-transformed data

Also the revised model, applied on the log-transformed data, seems to perform rather well on the monthly maximum data. The normal probability plots of the residuals suggest that the model revision is an improvement compared to the basic model, but the estimated losses are somewhat greater.

The expected contributions from the $\mu(x)$-field are between 1.76 and 1.95, but the interpretation is different. $e^{\mu(x)}$ is now a multiplicative factor for the monthly maximum significant wave height at location x, varying between 5.8 and 7.0 over the area. The mean

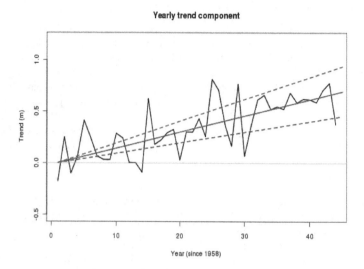

Figure 6. Estimated temporal trends of the North Atlantic wave climate; basic model

contributions from the space-time dynamic part, $\theta(x, t)$ corresponds to factors between 0.70 and 1.4 for different times and locations. Hence, this component contributes from -30% to +40%. The seasonal component corresponds to a factor 0.68 for calm seasons and 1.5 in rough seasons.

The mean estimated long-term trend from the linear model corresponds to a factor about 1.07 over the data-period. The 90% credible interval ranges from 1.03 to 1.12. For typical monthly maximum significant wave heights of, say, 5 and 8 meters respectively, this corresponds to expected increases of about 36 and 57 cm. However, for more extreme sea states, say significant wave heights of 10 or 15 meters respectively, corresponding expected increases would be 70 cm and more than 1 meter respectively. Overall, these trends are somewhat smaller than the trends estimated from the non-transformed data, but the trends pertaining to extreme conditions are greater. A QMLE-estimate for bias correction ([18]) due to retransformation has been adopted and is incorporated into the estimates above, see [39]. The estimated expected long-term trends with 90% credible interval are shown in figure 7 on the original, i.e. re-transformed scale.

Also the estimated trends obtained from the log-transformed data were extrapolated in order to obtain an estimate of future trends in the wave climate. Over 100 years, the expected future increase in monthly maximum wave height corresponds to a factor of 1.15, with a 95% credibility of a trend factor larger than 1.04. Assuming such trends to persist and valid for average monthly maximum sea states of 5 and 8 meters in calm and rough seasons, the expected increase is about 75 cm and 1.2 meters respectively. However, for more extreme sea states, with significant wave height of, say, 10 and 15 meters, expected increases would be 1.5 and 2.3 meters respectively towards the year 2100.

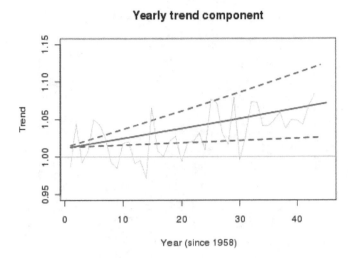

Figure 7. Estimated temporal trends of the North Atlantic wave climate; log-transformed model

6.3. Extended model with regression component

With the extended model, the stochastic relationship between significant wave height and atmospheric levels of CO_2 is exploited together with future projections of CO_2 trends in order to obtain estimates of future trends in the wave climate. Expectantly, the estimated trends should be similar to the trends estimated with the basic model, but future projections can be obtained from various CO_2 projections and may differ from the extrapolated linear trends.

It is noted that for the extended model, the long-term trend contribution does not necessarily start at 0 for t=0, but in the results presented herein, necessary adjustments have been made so that the long-term trend effectively starts at 0. For extracting the expected trends over the period 1958-2001, the long-term trend is adjusted to be 0 in 1958 and for the future projections towards 2100, the trend contribution is adjusted to start at 0 in 2001. This does not affect the relative trend between two points in time, but is accompanied by a similar but opposite adjustment of the mean spatial field.

The contributions from the adjusted time-independent field, $\mu(x)$, varies between 6.3 and 7.5 meters over the area, and this is in reasonable agreement to the estimates obtained from the basic model. The short-term dynamic contribution from $\theta(x,t)$ varies from -1.1 to 1.9 meters and the mean seasonal contributions lie between ± 2.66 meters, which also agrees well with the estimates obtained from the basic model.

The contribution from the long-term trend, possibly due to climate change, is shown in figure 8, corresponding to an expected increase of 59 cm over the period. This is somewhat lower than the estimated trend from the basic model, but still agrees fairly well. The 90% credible interval ranges from 16 to 92 cm increases in monthly maximum significant wave height.

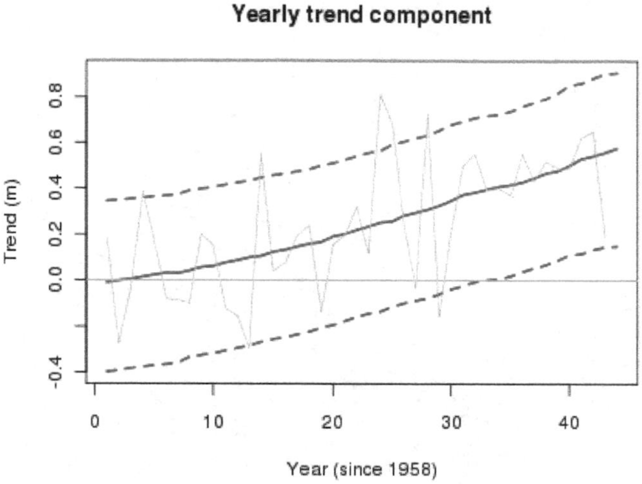

Yearly trend component

Figure 8. Estimated temporal trends of the North Atlantic wave climate; model with CO_2 regression

One of the main motivations for including the CO_2 regression component into the model was to facilitate future projections. Hence, projections of future significant wave heights are made from two future scenarios for CO_2 levels, referred to as the A2 and B1 scenario respectively. The corresponding projected trends of significant wave height are illustrated in figure 9, and it can be seen that scenario A2 yields future projections corresponding to an increase of 5.4 meters and the B1 scenario corresponds to an increase of 1.9 meters towards 2100 compared to the year 2001. The large difference between the two projections is due to the different CO_2 levels projected by the two scenarios. However, both the projected trends are considerably larger than the one obtained from extrapolating the linear trend obtained from the basic model.

The expected future projections including 90% credible intervals are illustrated in figure 10. The credible intervals are calculated from the credible intervals of the distribution of (γ, η) and do not include the uncertainty due to ε_T. For scenario A2, the 90% credible interval at year 2100 corresponds to increases in monthly maximum significant wave height over the 21 century ranging from 2.7 meters to 8.1 meters. For scenario B1, the corresponding credible interval covers a range between 1.2 to 2.6 meters increase from 2001 to 2100.

6.4. Model comparison

A crude comparison of the different model alternatives can be carried out by comparing the resulting posterior estimates of the model parameters (see [38–40]). By doing so, it is observed that the spatial features of the model seem to be barely affected by the model alterations. Since the model extensions were only related to the temporal trend, this is reassuring. The seasonal part of the model also seems to behave similarly over the model alternatives. Hence, the main

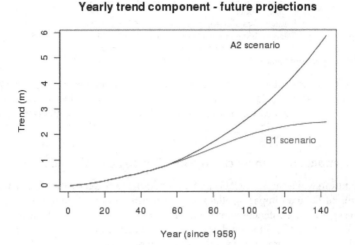

Figure 9. Estimated future trends in the North Atlantic wave climate for two emission scenarios

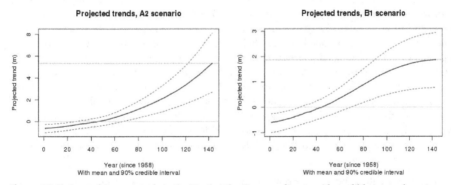

Figure 10. Estimated future trends in the North Atlantic wave climate with credible intervals

differences are, as would be expected, related to the long term temporal trend and the future projections.

The models may also be compared by way of the loss functions for short-term prediction. The estimated losses corresponding to the two loss functions for each of the model alternatives discussed herein are presented in table 2.

It is observed that both the models fitted to the original data are associated with lower losses compared to the model for log-transformed data. However, it is noted that comparison might not be fair for predictions made on log-transformed data so it does not necessarily mean that the revised model performs worst. Furthermore, the extended model with a CO_2 regression

Model alternative	L_s	L_w
Basic model	2.576	2.691
Revised model (log-transform)	3.346	3.412
Extended model (CO_2 regression)	2.562	2.674

Table 2. Model comparison: Standard and weighted loss functions

component yields the lowest losses, which indicates that this is an improvement compared to the basic model. The differences are small, however, and the estimated losses cannot be used to reliably distinguish between the models, which all seem to describe the data reasonably well.

7. Potential impact of climate change on ship structural loads

If indeed the future wave climate will become rougher, as predicted by the models presented herein, this might have an impact on the safety of maritime transportation, since ships are then likely to experience greater environmental loads. Extreme environmental loads represent a serious hazard to ship operations and any increasing trends might thus lead to higher risk, if not properly accounted for in design and operational procedures.

Having identified a trend in the significant wave height data, it would therefore be of great interest to consider how such results could be related to the calculation of future environmental loads and responses on ships and other floating structures. Trends in the ocean wave climate will obviously also be important for offshore and coastal structures, and the results can generally be applied also to offshore and coastal structures design. However, if applied to fixed installations, location specific data should be used; North Atlantic data are used only for ship design. For the purpose of this study, the trends towards 2100 estimated from the basic model and the extended model with a CO_2 regression component and scenario B1 will be assumed. It will be investigated how to relate such trends to the calculations of ship structural loads and responses. It is emphasized that potential influence of such trends on structural design, as was discussed in [8] is not considered explicitly herein. Results pertaining to any other projection period, such as 30 or 40 years ahead in time, could also easily have been used.

The trends estimated above correspond to an addition, 100 years ahead in time, with mean 1.6 meters and standard deviation 0.39 meters from the basic model and mean 1.9 meters and standard deviation 0.65 meters from the extended model adopting the B1 scenario. The mean and standard deviation of the climatic trend contribution will be denoted by μ_{ct} and σ_{ct} when stemming from the basic model and μ_{B1} and σ_{B1} when estimated by the extended model with scenario B1, respectively, i.e. an additive trend, $T \sim (\mu., \sigma.^2)$ will be assumed.

$$\mu_{ct} = 1.6m \qquad \sigma_{ct} = 0.39m$$
$$\mu_{B1} = 1.9m \qquad \sigma_{B1} = 0.65m \tag{17}$$

It is noted that the climate trend is estimated from monthly maxima although it is applied to the whole body of the H_s distribution (the marginal distribution of significant wave height). Thus the revised H_s distribution is more representative for high values of H_s. When the impact of the trend is explored, extreme loads are considered and this makes these assumptions less troublesome; this simplification is considered acceptable for extremes but neither for fatigue calculations nor specification of operational criteria when lower sea states are of importance.

7.1. Conditional modelling of joint metocean parameters

The marginal distribution of significant wave height is normally not sufficient for load and response calculations of marine structures; the joint distribution of several metocean parameters is required. As a minimum, the joint distribution of significant wave height and wave period is needed.

The above trends were extracted from the corrected ERA-40 data (C-ERA-40) over an area in the North Atlantic. Due to lack of information about wave period in the C-ERA-40 data, the joint distribution of significant wave height and wave period used for load calculations are based on the ERA interim data set[3] for a particular location. However, that location is contained within the area considered by the C-ERA-40 data and is assumed representative for the whole area. Furthermore, main features of the C-ERA-40 and ERA$_{\text{Interim}}$ data sets are similar, and it is assumed that any bias would be similar in the two data sets. The long-term trends obtained in the present study are therefore incorporated in the established joint distribution of significant wave height and wave period based on the ERA$_{\text{interim}}$ data.

It has previously been proposed to model the marginal distribution of significant wave height, H_s, according to a 3-parameter Weibull distribution and the conditional distribution of the wave period, T, conditional on the significant wave height, as a log-normal distribution ([4, 26]). Hence, the joint distribution of significant wave height and wave period will be the product of a Weibull and a log-normal distribution given H_s (eq. 18) according to the Conditional Modelling Approach (for several met-ocean parameters see [2, 3]). The 3-parameter Weibull distribution was first applied to describe significant wave height by [29].

$$f_{H_s,T}(h,t) = f_{H_s}(h) f_{T|H_s}(t|h) \tag{18}$$

It is assumed that the trend in the significant wave height corresponds to a modified marginal distribution for the significant wave height, but that the distribution of wave period, conditional on the significant wave height, remains unchanged. It is noted that even though the conditional distribution is assumed unchanged, the marginal distribution of the wave period will obviously change, so this assumption seems reasonable.

The 3-parameter Weibull distribution is parametrized by the parameters γ (location), α (scale) and β (shape), as shown in eq. 19.

$$f(x) = \frac{\beta}{\alpha} \left(\frac{x-\gamma}{\alpha} \right)^{\beta-1} e^{-\left(\frac{x-\gamma}{\alpha}\right)^{\beta}} \qquad\qquad x \geq \gamma \tag{19}$$

It is assumed that the distribution of the significant wave height after the trend has been added can be approximated by a 3-parameter Weibull distribution with the same shape parameter, i.e. that the trend can be modelled as a modification of the location and scale parameters of

---

[3] Website: http://www.ecmwf.int/research/era/do/get/era-interim

the 3-parameter Weibull distribution. A simulation study confirms that this is a reasonable approximation. With these assumptions, the modified parameters due to the long-term trend can be calculated so that the modified Weibull distribution has the correct expectation and variance, resulting in the modified parameters in eqs. 20-21.

$$\gamma \to \gamma' = \gamma + \mu_{ct} + \Gamma \left(\frac{1}{\beta} + 1 \right) \left[\alpha - \sqrt{\alpha^2 + \frac{\sigma_{ct}^2}{\Gamma \left(\frac{2}{\beta} + 1 \right) - \Gamma \left(\frac{1}{\beta} + 1 \right)^2}} \right] \tag{20}$$

$$\alpha \to \alpha' = \sqrt{\alpha^2 + \frac{\sigma_{ct}^2}{\Gamma \left(\frac{2}{\beta} + 1 \right) - \Gamma \left(\frac{1}{\beta} + 1 \right)^2}} \tag{21}$$

The 3-parameter Weibull distribution was fitted to significant wave height data for one location from the ERA-40$_{\text{Interim}}$ data and the estimated parameters together with the modified parameters as a result of adding the projected long-term trends (over 100 years) are given in table 3. The corresponding mean and standard deviation of the distributions are also given.

	α	β	γ	$E[H_s]$	$sd[H_s]$
Fitted distribution	2.776	1.471	0.8888	3.401	1.737
Modified parameters (Basic model trend)	2.846	1.471	2.393	4.969	1.781
Modified parameters (Extended model / B1)	2.965	1.471	2.613	5.296	1.855

Table 3. Fitted and modified parameters for the 3-parameter Weibull distribution for significant wave height

It is observed that the mean of the modified distribution is changed quite drastically, whereas there is only a slight increase in the standard deviation as a result of adding the climatic trend with uncertainties.

The conditional distribution of wave period is modelled as a log-normal distribution where the parameters are modelled as functions of significant wave height, as shown in eqs. 22-23. By assumption, this conditional distribution is not expected to change due to climatic trends, and the parameters a_i and b_i for $i = 1, 2, 3$ are estimated from the data. The resulting joint densities of the original and the modified distributions for significant wave height, H_s, and zero-up-crossing period, T_z, are illustrated by the contour plots in figure 11 (on the same scale). It is noted that T_z is one of several ways of describing the wave period, T.

$$\mu_t = E[\ln T_z | H_s = h_s] = a_1 + a_2 h_s^{a_3} \tag{22}$$
$$\sigma_t = sd[\ln T_z | H_s = h_s] = b_1 + b_2 e^{b_3 h_s} \tag{23}$$

7.2. Case study: Impact of long-term trends on the load assessment of an oil tanker

As an illustrative example, load characteristics will be calculated for an oil tanker of 250 m length and 40 m width with the same characteristics as the one reported in [5].

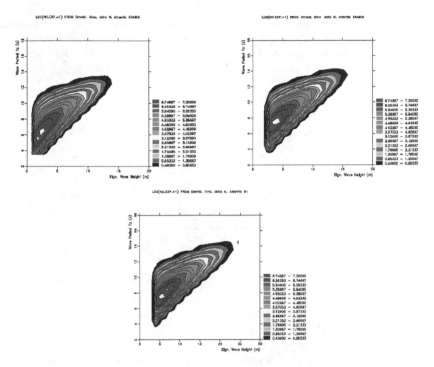

Figure 11. Contour plots of the joint distribution of significant wave height and zero-up-crossing period; Fitted distributions (top left) and modified distributions with trends estimated from the basic model (top right) and the extended model with the B1 scenario (bottom)

When specifying design criteria as well as carrying out load and response assessment for marine structures a full long-term load and response analysis can be applied, or alternatively, the environmental contour concept outlined in [48] can be used (IFORM). The latter is a valid, simplified and rational method of estimating extreme conditions and is recommended by DNV ([16]). The idea is to define contours in the environmental parameter space (usually H_s, T_z) within which extreme responses with a given return period should lie. It requires determination of the joint environmental model of sea state variables of interest. It should be noticed that the contours are found by relating sea state variables to the standard normal variables, an assumption that may affect their accuracy. Furthermore, adding the trend introduces a dependency between the sea states at subsequent times, but the effect this might have on the return values have been ignored in this study. Presumably, since the variability of the estimated trend is small in comparison to the variability of sea states, this effect is not very great.

Figure 12 shows the environmental contour lines of H_s and T_z for the North Atlantic location considered in the present study. The 1, 10, and 25-year return period levels calculated by IFORM are shown for the fit to the original ERA$_{Interim}$ data and for the corrected fits where the long-term trends are included. The 3-parameter Weibull distributions for H_s given in table 3 and the conditional log-normal distribution for T_z have been used in the analysis. As expected, the modification of the distribution for significant wave height moves the environmental contours upwards and to the right. Furthermore, the long-term trend correction has narrowed the contours and increased the maximum 1, 10 and 25-year return H_s and related T_z.

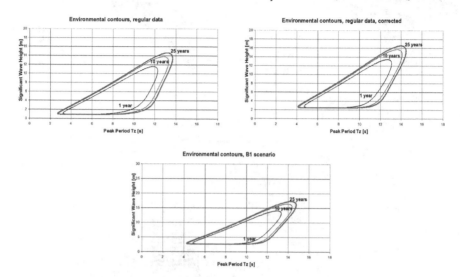

Figure 12. Environmental contour lines for the $ERA_{Interim}$ data derived from the original data (top left) and modified with the climatic trend over 100 years from the basic model (top right) and from the extended model with the B1 scenario (bottom)

The 25-year stress amplitudes for the considered oil tanker have been calculated in the 25-year sea states (H_s, T_z) given by the environmental contour lines. A 3-hour sea state duration and a Rayleigh distributed stress process in a short-term sea state (see [5]) have been assumed in the calculations. Table 4 includes the results of the analysis for the original joint (H_s, T_z) fit and the modified ones, taking the estimated 100-year long-term trends into account. The response characteristics obtained using the original (H_s, T_z) fit to the $ERA_{Interim}$ data are referred to herein as a Base Case and only relative increases in comparison to the Base Case are given in table 4.

	Stress amplitude (MPa)	Response period (s)
Base case	1.0	1.0
Modified fit - Basic model	1.07	1.02
Modified fit - Extended model / B1	1.10	1.02

Table 4. 25-year extreme load characteristics

As seen in table 4 incorporation of the long-term trend in the H_s distribution has increased the 25-year stress amplitude significantly and also the zero-crossing response period has been increased. The 25-year stress amplitude has increased by 7% or 10% while the zero-crossing response period has increased by 2% due to the estimated long-term trend over 100 years. It is noted that similar calculations have not been done for the A2 scenario, but the effect would presumably be even larger for such a worst-case trend. Furthermore, the potential effect of the modified environmental contours on the structural loads is highly ship-dependent and even though the loads were found to increase significantly for this particular ship, it does not necessarily generalize to all types and sizes of ships.

8. Discussion

The models presented in this chapter aim at modelling the effect of climate change on the North Atlantic wave climate, and all model variations agree that the ocean wave climate has become rougher and is likely to become even rougher towards the next century. In [41], the models have been fitted to data for 11 other ocean areas around the globe with increasing trends predicted also for 9 of these.

Different emission scenarios have been assumed in the extended model to obtain different future projections of the wave climate, and it has been seen that adopting an extreme emission scenario, such as the A2 scenario, corresponds to predicting extreme future trends in significant wave height. With a more moderate emission scenario, such as B1, the resulting future projections are still larger than the extrapolation of observed recent trends, as predicted by the basic model. As for any future climate predictions, uncertainties are large and it is difficult to determine which predictions are best. According to the two loss functions utilized in this study (eqs. 13-14), the extended model seems to represent an improvement in describing the data at hand but this does not necessarily mean that the projections from this model are more reliable than the others; significant uncertainties are also related to the CO_2 scenarios proposed by IPCC. However, the extended model may adopt different CO_2 scenarios and investigate their effects on the trends in the future ocean wave climate.

An implicit assumption inherent in the extended model is that there is stochastic dependence between atmospheric CO_2 levels and the ocean wave climate, and also that this relationship will remain essentially the same in the future. It is assumed that an increase of greenhouse gases in the atmosphere will increase the temperature and put more energy into the weather systems, leading to more powerful storms and wind fields. This might again change the ocean wave climate, since it is well known that ocean waves are generated by wind and air pressure gradients. This is obviously a simplification, and it is possible to refine the model with different layers of dynamics and relationships, e.g. including projections of wind or pressure fields as explanatory variables in the model. Notwithstanding, it is argued that the physics underlying the wave generation process is inevitably implicit in the data, and when applying the models to a particular data set any such physical effects would unavoidably be incorporated by way of the data. On the other hand, the stochastic models are affected by data uncertainties and possible biases.

Comparing the projections obtained with the different Bayesian hierarchical space-time models with previous studies, it is seen that apart from the predictions pertaining to the A2 scenario, the projections are comparable to those made for the North Atlantic in e.g. [12, 14, 15, 19, 42-44]. The uncertainties are large, the estimated 90% credible intervals correspond to about $\pm$ 50% of the expected projections, and the intervals generally overlap. It should also be kept in mind that the trends predicted herein pertain to the monthly maxima and the maxima might experience a greater change than moderate sea states, as also suggested in e.g. [49]. The A2 scenario can be construed as a worst case scenario and is important to consider from a precautionary perspective. Nevertheless, even stronger trends were predicted in [31], albeit for an area in the Pacific ocean.

It should be stressed that even though the models detect trends in the data, it does not necessarily mean that the trend is a direct consequence of anthropogenic climate change. It might be a result of decadal natural variability, as discussed in e.g. [9], and wave climate variability has been reported to be considerable on different temporal scales ([17]). Great care should therefore be taken when interpreting the meaning and the origin of this trend, even though the correlation between anthropogenic CO_2 emissions and the wave climate are found to be strong in this study.

One practical implication of the predicted changes in the ocean wave climate due to global climate change is that the structural loads and stresses on ships and other marine structures might increase notably in the future. A case study indicates increases of up to 10% over the current century. This is not negligible, and it is therefore recommended to carefully consider and take into account the potential impact of climate change in the design and construction of ocean going ships to avoid jeopardizing the safety of future maritime operations.

9. Conclusions

This chapter has been concerned with the potential impact of global climate change on the ocean wave climate and, consequently, on the risk of maritime transportation. A Bayesian hierarchical space-time model that has been developed to model the effect of climate change on the ocean wave climate has been presented. Different versions of the model have been discussed and they all agree in a non-negligible positive trend in the monthly maximum significant wave height over a selected area of the North Atlantic ocean. Estimated expected additive trends towards 2100 range from 1.6 - 5.4 meters and expected multiplicative trends in the order of 15% are predicted. Assuming an average monthly maximum significant wave height of 7.5 meters, the estimated trends related to the B1 and A2 scenarios correspond to centurial increases of 25% - 72%, which are indeed significant. However, the uncertainties are large, and 90% credible intervals for the expected trends range from 1.2 to 2.6 meters for the B1 scenario and 2.7 to 8.1 meters for the A2 scenario.

One of the advantages of using a stochastic model is that estimates of the uncertainties are given explicitly. These are important when future projections are to be incorporated in risk analyses or utilized in probabilistic load calculations as illustrated by an example in this chapter. The case study reveals that the effect of the predicted trend in the ocean wave climate on environmental loads of ships is far from being negligible, and that this may need to be taken into account in design and construction of ships. Obviously, a roughening of the ocean wave climate also has the potential to severely impact other areas of society as well, related to maritime, offshore and coastal activities. Combined with sea level rise and other possible effects of climate change, coastal areas throughout the globe may be seriously affected.

How to adapt to climate change is one of the most important questions in society today. It is a political question and perhaps a moral question as much as it is a scientific question.   Nevertheless, an important prerequisite for making well-founded decisions is reliable predictions of the future effect of climate change.   The stochastic model presented in this chapter aims at contributing to this discussion by providing a model for predicting the effect of climate change on the ocean wave climate.   Such an effect could again have practical implications on many areas of society, most notably related to marine and coastal management.   It is acknowledged that the models represent a simplification of reality, as inevitably all models do, and that there is potential for improvements to the models. Nonetheless, it is believed that the study presented herein is an important contribution to the scientific debate on the effects of climate change, and it is a hope that it can spur further debate and motivate further research into the effects of climate change on the future ocean wave climate.

Acknowledgments

The authors want to express their thanks to Dr. Andreas Sterl at KNMI for kindly providing the significant wave height data used in this analysis and for clarifying some issues discovered when investigating the data. Thanks also to Dr. Pieter Tans for kind permission to use the NOAA ESRL CO_2-data. Furthermore, thanks to the Norwegian Meteorological Institute for providing the $ERA_{Interim}$ data. The data on future projections of atmospheric concentration of CO_2 were obtained from the IPCC.

Author details

Erik Vanem, Bent Natvig and Arne Bang Huseby
University of Oslo, Norway

Elzbieta M. Bitner-Gregersen
Det Norske Veritas, Norway

10. References

[1] Arena, F. & Pavone, D. [2009].   A generalized approach for long-term modelling of extreme crest-to-trough wave heights, *Ocean Modelling* 26: 217–225.

[2] Bitner-Gregersen, E. & Haver, S. [1989]. Joint long term description of environmental parameters for structural response calculation, *Proc. 2nd International Workshop on Wave Hindcasting and Forecasting.*

[3] Bitner-Gregersen, E. & Haver, S. [1991].   Joint environmental model for reliability calculations, *Proc. 1st International Offshore and Polar Engineering conference (ISOPE 1991),* The International Society of Offshore and Polar Engineering (ISOPE).

[4] Bitner-Gregersen, E. M. [1988].   Appendix: Joint long term distribution of Hs, Tp, *Probabilistic Calculation of Design Criteria for Ultimate Tether Capacity of Snorre TLP,* Madsen, H.O., Rooney, P. and Bitner-Gregersen, E. Det Norske Veritas Report No. 87-31.

[5] Bitner-Gregersen, E. M., Cramer, E. H. & Løseth, R. [1995].   Uncertainties of load characteristics and fatigue damage of ship structures, *Marine Structures* 8: 97–117.

[6] Bitner-Gregersen, E. M. & de Valk, C. [2008].   Quality control issues in estimating wave climate from hindcast and satellite data, *Proc. 27th International Conference on*

Offshore Mechanics and Arctic Engineering (OMAE 2008), American Society of Mechanical Engineers (ASME).

[7] Bitner-Gregersen, E. M. & Hagen, Ø. [1990].   Uncertainties in data for the offshore environment, *Structural Safety* 7: 11–34.

[8] Bitner-Gregersen, E. M., Hørte, T. & Skjong, R. [2011].   Potential impact of climate change on tanker design, *Proc. 30th International Conference on Ocean, Offshore and Arctic Engineering (OMAE 2011)*, American Society of Mechanical Engineers (ASME).

[9] Caires, S. & Sterl, A. [2005a]. 100-year return value estimates for ocean wind speed and significant wave height from the ERA-40 data, *Journal of Climate* 18: 1032–1048.

[10] Caires, S. & Sterl, A. [2005b].   A new nonparametric method to correct model data: Application to significant wave height from ERA-40 re-analysis, *Journal of Atmospheric and Oceanic Technology* 22: 443 – 459.

[11] Caires, S. & Swail, V. [2004]. Global wave climate trend and variability analysis, *Preprints of 8th International Workshop on Wave Hindcasting and Forecasting*.

[12] Caires, S., Swail, V. R. & Wang, X. L. [2006].   Projection and analysis of extreme wave climate, *Journal of Climate* 19: 5581–5605.

[13] Cressie, N. & Wikle, C. K. [2011]. *Statistics for Spatio-Temporal Data*, Wiley.

[14] Debernard, J. B. & Røed, L. P. [2008]. Future wind, wave and storm surge climate in the Northern Seas: a revisit, *Tellus* 60A: 427–438.

[15] Debernard, J., Sætra, Ø. & Røed, L. P. [2002]. Future wind, wave and storm surge climate in the northern North Atlantic, *Climate Research* 23: 39 – 49.

[16] DNV [2010].   *Environmental Conditions and Environmental Loads*, Det Norske Veritas. DNV-RP-C205.

[17] Dodet, G., Bertin, X. & Tabora, R. [2010].   Wave climate variability in the North-East Atlantic Ocean over the last six decades, *Ocean Modelling* 31: 120–131.

[18] Ferguson, R. [1986].   River loads underestimated by rating curves, *Water Resources Research* 22: 74–76.

[19] Grabemann, I. & Weisse, R. [2008]. Climate change impact on extreme wave conditions in the North Sea: an ensemble study, *Ocean Dynamics* 58: 199 – 212.

[20] IPCC [2001]. *Climate Change 2001: The Scientific Basis*, Cambridge University Press.

[21] IPCC [2007a]. Climate change 2007: Synthesis report, *Technical report*, Intergovernmental Panel on Climate Change.

[22] IPCC [2007b].   *Climate Change 2007: The Physical Sciences Basis. Contribution of Working Group I to the Fourth Assessment Report of the Intergovernmental Panel on Climate Change*, Cambridge University Press.

[23] IPCC [2012]. *Managing the Risks of Extreme Events and Disasters to Advance Climate Change Adaptation*, Cambridge University Press.

[24] Jain, A. K., Kheshgi, H. S. & Wuebbles, D. J. [1994].   Integrated science model for assessment of climate change, *Technical Report UCRL-JC-116526*, Lawrence Livermore National Laboratory.

[25] Kheshgi, H. S. & Jain, A. K. [2003].   Projecting future climate change: Implications of carbon cycle model intercomparisons, *Global Biogeochemical Cycles* 17: 1047.

[26] Mathisen, J. & Bitner-Gregersen, E. [1990]. Joint distributions for significant wave height and wave zero-up-crossing period, *Applied Ocean Research* 12: 93–103.

[27] Nakićenović, N., Alcamo, J., Davis, G., de Vries, B., Fenhann, J., Gaffin, S., Gregory, K., Grügler, A., Jung, T. Y., Kram, T., La Rovere, E. L., Michaelis, L., Mori, S., Morita, T., Pepper, W., Pitcher, H., Price, L., Riahi, K., Roehrl, A., Rogner, H.-H., Sankovski, A., Schlesinger, M., Shukla, P., Smith, S., Swart, R., van Rooijen, S., Victor, N. & Dadi, Z. [2000]. *Emissions scenarios*, Cambridge University Press.

[28] Natvig, B. & Tvete, I. F. [2007]. Bayesian hierarchical space-time modeling of earthquake data, *Methodology and Computing in Applied Probability* 9: 89–114.

[29] Nordenstrøm, N. [1973]. A method to predict long-term distributions of waves and wave-induced motions and loads on ships and other floating structures, *Technical Report 81*, Det Norske Veritas.

[30] Robert, C. P. & Casella, G. [2004]. *Monte Carlo Statistical Methods*, second edn, Springer.

[31] Ruggiero, P., Komar, P. D. & Allan, J. C. [2010]. Increasing wave heights and extreme value projections: The wave climate of the U.S. Pacific Northwest, *Coastal Engineering* 57: 539–552.

[32] Sterl, A. & Caires, S. [2005]. Climatology, variability and extrema of ocean waves: The web-based KNMI/ERA-40 wave atlas, *International Journal of Climatology* 25: 963–977.

[33] Thoning, K. W., Tans, P. P. & Komhyr, W. D. [1989]. Atmospheric carbon dioxide at Mauna Loa observatory 2. analysis of the NOAA GMCC data, 1974-1985, *Journal of Geophysical Research* 94: 8549–8565.

[34] Uppala, S. M., Kållberg, P. W., Simmons, A. J., Andrae, U., Da Costa Bechtold, V., Fiorino, M., Gibson, J. K., Haseler, J., Hernandez, A., Kelly, G. A., Li, X., Onogi, K., Saarinen, S., Sokka, N., Allan, R. P., Andersson, E., Arpe, K., Balmaseda, M. A., Beljaars, A. C. M., Van de Berg, L., Bidlot, J., Bormann, N., Caires, S., Chevallier, F., Dethof, A., Dragosavac, M., Fisher, M., Fuentes, M., Hagemann, S., Hólm, E. Hoskins, B. J., Isaksen, L., Janssen, P. A. E. M., Jenne, R., McNally, A. P., Mahfouf, J.-F., Morcrette, J.-J., Rayner, N. A., Saunders, R. W., Simon, P., Sterl, A., Trenberth, K. E., Untch, A., Vasiljevic, D., Vitebro, P. & Woolen, J. [2005]. The ERA-40 re-analysis, *Quarterly Journal of the Royal Meteorological Society* 131: 2961–3012.

[35] Vanem, E. [2011]. Long-term time-dependent stochastic modelling of extreme waves, *Stochastic Environmental Research and Risk Assessment* 25: 185–209.

[36] Vanem, E. & Bitner-Gregersen, E. [2012]. Stochastic modelling of long-term trends in the wave climate and its potential impact on ship structural loads, *Applied Ocean Research* 37: 235–248.

[37] Vanem, E., Huseby, A. B. & Natvig, B. [2012a]. A Bayesian hierarchical spatio-temporal model for significant wave height in the North Atlantic, *Stochastic Environmental Research and Risk Assessment* 26: 609–632.

[38] Vanem, E., Huseby, A. B. & Natvig, B. [2012b]. Bayesian hierarchical spatio-temporal modelling of trends and future projections in the ocean wave climate with a CO_2 regression component, submitted.

[39] Vanem, E., Huseby, A. B. & Natvig, B. [2012c]. Modeling ocean wave climate with a Bayesian hierarchical space-time model and a log-transform of the data, *Ocean Dynamics* 62: 355–375.

[40] Vanem, E., Huseby, A. B. & Natvig, B. [2012d]. A stochastic model in space and time for monthly maximum significant wave height, *Proc. Ninth International Geostatistics Congress (Geostats 2012)*: 505–517.

[41] Vanem, E., Natvig, B. & Huseby, A. B. [2012]. Modelling the effect of climate change on the wave climate of the world's oceans, *Ocean Science Journal* 47: 123–145.

[42] Wang, X. J., Zwiers, F. W. & Swail, V. R. [2004]. North Atlantic ocean wave climate change scenarios for the twenty-first century, *Journal of Climate* 17: 2368–2383.

[43] Wang, X. L. & Swail, V. R. [2006a]. Climate change signal and uncertainty in projections of ocean wave heights, *Climate Dynamics* 26: 109–126.

[44] Wang, X. L. & Swail, V. R. [2006b]. Historical and possible future changes of wave heights in northern hemisphere oceanc, *in* W. Perrie (ed.), *Atmosphere-Ocean Interactions*, Vol. 2 of *Advances in Fluid Mechanics, Vol. 39*, WIT Press, chapter 8, pp. 185–218.

[45] Wikle, C. K. [2003]. Hierarchical models in environmental science, *International Statistical Review* 71: 181 – 199.

[46] Wikle, C. K., Berliner, L. M. & Cressie, N. [1998]. Hierarchical Bayesian space-time models, *Environmental and Ecological Statistics* 5: 117 – 154.

[47] Wikle, C. K., Milliff, R. F., Nychka, D. & Berliner, L. M. [2001]. Spatiotemporal hierarchical Bayesian modeling: Tropical ocean surface winds, *Journal of the American Statistical Association* 96: 382 – 397.

[48] Winterstein, S., Ude, T., Cornell, C., Bjerager, P. & Haver, S. [1993]. Environmental parameters for extreme response: Inverse FORM with omission factors, *Proc. 6th International Conference on Structural Safety and Reliability*.

[49] Young, I., Zieger, S. & Babanin, A. [2011]. Global trends in wind speed and wave height, *Science* 332: 451–455.

Chapter 9

A Study of Climate Change and Cost Effective Mitigation of the Baltic Sea Eutrophication

Martin Lindkvist, Ing-Marie Gren and Katarina Elofsson

Additional information is available at the end of the chapter

1. Introduction

Eutrophication of coastal marine waters is globally considered to be a serious environmental problem [1, 2]. The Baltic Sea is the world's largest brackish-sea and damages from eutrophication have been documented since the early 1960s by a large number of different studies [e.g. 3, 4]. In response to eutrophication of the sea the riparian states formed the administrative body HELCOM in charge of policies for improving the Baltic Sea and entered ministerial agreements on nutrient reduction in 1988 and 2007. Although nutrient reductions have been made, the 50 percent reduction agreed upon in 1988 has been far from reached and the ecological status of the sea continues to deteriorate. In order to reach the ecological goal of "clear water", which is one main objective of the 2007 treaty, large reductions of both phosphorous and nitrogen are necessary. The cost of these nutrient reductions can be substantial, not the least since many low cost abatement options have already been implemented. In this respect it is important to evaluate if and how future nutrient loads will change and how this will affect costs for achieving stipulated targets.

Climate change and structural changes in the agricultural sector are considered to be the major drivers of future nutrient loads to the Baltic Sea [5]. Climate change is expected to change the precipitation pattern in the drainage basin. This is expected to lead to an increase in mean annual river-flows in the northen drainage basins of the Sea and a decrease in mean annual river-flows in the southern part of the catchment [6, 7]. Changes in run-off and river flows explain 71-97 percent of the variability in land-sea fluxes of nutrients [6]. Climate change will therefore affect the magnitude of future nutrient loads to the Baltic Sea. The purpose of this paper is to calculate cost-effective solutions to reductions of nutrient loads stipulated by the Baltic Sea Action Plan (BSAP) [6] under different scenarios with respect to impacts of climate change on nutrient loads. Since climate change is not occurring in a vacuum we will also take the effect of agricultural change and demographic changes on future nutrient loads into consideration. This is carried out by means of a numerical dynamic discrete model of control costs for abatement in the riparian countries of the Baltic Sea.

Similar to several other international water bodies, the Baltic Sea contains a number of interlinked and heterogeneous marine basins. The ecosystem conditions in these basins differ, and the BSAP therefore suggests different nutrient load targets for the basins. However, since the basins are coupled, nutrient load reduction to one basin affects all the other basins. This means that both dynamic and spatial distribution of abatement need to be taken into account when identifying cost effective timing and location of nutrient abatement. Starting in mid 1990s there is by now a relatively large economics literature on cost effective or efficient nutrient load reductions to the Baltic Sea e.g. [8-21], but most of these studies calculate cost effective or efficient allocation of abatement among the riparian countries in a static setting [8-10, 12-14, 17, 21].

A majority of the few studies accounting for nutrient dynamics considers only one marine basin, disregards the heterogeneity among marine basins, and/or restrict the number of nutrient related activities [15, 16, 10, 18, 19]. The focus is often on optimal nutrient management in one drainage basin including only agriculture [15, 16] or this sector together with sewage treatment [10, 18, 19]. The only study covering the entire drainage basin of the Baltic Sea, accounting for coupled heterogeneous marine basins with respect to dynamics of both nitrogen and phosphorus, and including several nutrient-emitting sectors is [20]. However, none of the studies applied on eutrophication in the Baltic Sea or in lakes evaluate consequences on optimal cost paths under different climate changes scenarios and future development with respect to demography and agriculture. In [21] they addresses the same question as in this paper i.e. impacts of climate change on cost effective management of eutrophied water, but applies a static approach to a sub drainage basin of the Baltic drainage basin. In order to calculate impacts on total abatements costs and associated allocation among the riparian countries the dynamic model in [20] is developed to account for the different scenarios of future nutrient loads. This paper therefore extends earlier literature on dynamic management of eutrophied seas and lakes by adding scenario analysis of climate change to the spatial and temporal perspectives on cost effective nutrient management.

The paper is organised as follows. First the numerical model underlying the calculations of effective solutions is presented, i.e. the allocation of abatement among drainage basins and during time which minimizes total cost for achieving nutrient load targets within a specific time period. Derivation of the climate change scenarios is carried out in section 3. In section 4 the cost effective achievement of the BSAP under different scenarios is presented and the paper ends with some tentative conclusions.

2. A brief presentation of the numerical model of dynamic and spatial nutrient management

The discrete dynamic model builds on [20], but adds a climate change dimension by alterations in business as usual (BAU) nutrient loads from different future changes in nutrient load; *i)* climate change induced impacts on nutrient leaching from given emission sources in the drainage basins, *ii)* the development of nutrient emissions from agriculture and *iii)* demographic impacts on nutrient loads. A scenario is then defined as a combination of these types of causes for changes in nutrient loads to the Baltic Sea, which are further described in Section 3.

The effect of any climate change scenario is calculated as the impacts of minimum costs for achieving predetermined nutrient concentration targets in the future compared to the reference case. The basis for target setting is the most recent ministerial agreement on nutrient load restrictions to the different marine basins presented in the HELCOM Baltic Sea Action Plan (BSAP) [22, 23], which are shown in Table 1. Since the targets are determined for marine basins and costs of nutrient load reductions are born by the nutrient emitting sectors in the drainage basins of the Baltic Sea four connected but different spatial layers are included in the numerical model; sub-catchments (24), countries (9), marine basins, (7) and the entire catchment (see Figure A1 in the appendix). The dynamic scale is captured by the responses to nutrient loads in each marine basin. A simplification is thus made by disregarding the dynamics of nutrient transports in the drainage sub catchment. The reason is the lack of harmonized data on nutrient dynamics for all sub-catchments and for both nitrogen and phosphorus. Such data is available only for the dynamics in the marine basins [24] and for nutrient transports between marine basins. On the other hand, there exist no quantifications of climate change on nutrient dynamics in the sea but only on nutrient transports from the catchments to the marine basins. It is therefore assumed that the climate change impacts on nutrient loads can be described as a proportional changes in nutrient loads from the emission sources in the reference case, see Appendix A. For given emissions at the sources such as agriculture land, climate change affects leaching and transports of nutrients to the sea which can either increase or decrease loads to the sea.

Based on simple analysis of the model accounting for the four different spatial layers in the Baltic Sea and its catchment it is shown that the net effect of climate change on costs for achieving predetermined nutrient concentration targets in different marine basins are determined by two counteracting factors: change in target stringency through impacts on the BAU loads in the reference scenario and the effects of abatement on nutrient loads (see Appendix A). If we consider only the impacts on the BAU loads proportional increases (decreases) in nutrient loads will increase (decrease) costs for achieving an unchanged nutrient concentration target compared with the reference case. On the other hand, higher nutrient loads also imply larger impact on nutrient loads from given abatement by a measure, which lowers costs of achieving the given targets. The combined net impact of these counteracting factors on the resulting abatement costs can be determined only by numerical analysis, which is carried out in Section 4.

Minimum costs are calculated by means of a dynamic discrete optimisation model including the four spatial layers, the structure of which is presented in Appendix A. Data on nutrient loads from the drainage basins, transports among marine basins, nutrient pools and dynamics in each marine basin are obtained from [20], and constitute the reference scenario. Gren et al. [20] applies an oceanographic model with transport among basins described by coefficient matrixes (Tables A3 and A4 in [20]) for calculating carry over rates of nutrient among periods (Table 1 in [20]), nutrient pools and concentration. The oceanographic model allows for individual description and simulation of different forms of nutrients occurring in the sea: inorganic, labile organic and refractory organic fractions. Inorganic and labile organic fractions are together considered as biologically available fraction that mainly determines eutrophication and is readily affected by human activity, while dynamics of refractory organic compounds driven mainly by natural processes are hardly significant for eutrophication. Nutrient loads, pools and concentrations in the reference case are therefore expressed in terms of bio-available fractions, see Table 1.

	Nutrient load, kton/year[1]		Nutrient pools, kton[2]		Nutrient concentration μM [3], reference		Nutrient concentration μM [3], target,	
	N	P	N	P	N	P	N	P
Bothnian Bay	25	2.5	183	7.4	8.73	0.16	9.93	0.15
Bothnian Sea	36	2.3	457	71.2	6.67	0.47	7.43	0.34
Baltic Proper	333	17.8	1330	435	7.31	1.08	6.28	0.55
Gulf of Finland	73	6.3	143	25.9	9.29	0.76	9.36	0.51
Gulf of Riga	61	2.1	86	12.7	14.51	0.97	22.81	0.64
Danish Straits	59	1.3	34	6.7	8.50	0.75	7.30	0.51
Kattegat	70	1.5	55	8.7	9.14	0.65	8.42	0.57
Total	657	33.8	2288	567				

1. Table B1 in Appendix B; 2. [20] Table 1; 3. [20] Table 5

Table 1. Bio-available nutrient loads/year, pools, and concentrations in the reference and target cases, N=nitrogen, P=phosphorus

The Baltic Proper receives that largest loads of nutrients every year and contains the largest pools of both nutrients. It also faces the relatively most stringent phosphorus reduction target; the concentration needs to be reduced by approximately 50 per cent. This is in contrast with the nitrogen concentration targets, which are more close to or even above those in the reference case. It can also be noticed that one country, Poland, accounts for 38 per cent of total phosphorus loads and for 30 percent of total nitrogen loads, see Table B1 in Appendix B.

Costs of nutrient load abatement are estimated by means of a pseudo data approach (see e.g. [25]). Unlike traditional sources, such data sets are not constrained by historical variations in, for example, factor prices and yields from land affecting land prices. Observations on costs and nutrient reductions are then obtained by using the static optimisation model in [26] for calculating minimum cost solutions for different levels of nutrient reductions to the coastal waters from each drainage basin. The static model in [26] contains a number of different measures for reducing water and airborne nitrogen and phosphorous loads from agriculture, industry and sewage. In total, the static model includes 14 measures affecting nitrogen loads and 12 measures changing phosphorous loads. These measures can be divided into three main categories: reductions in nutrients at the source, reductions in leaching of nutrients into soil and water for given nutrient emission levels, and reductions in discharges into the Baltic Sea for given emissions at sources and leaching into soil and water. Examples of the first class of measures include, among others, reductions in nitrogen fertilizers and reductions in livestock. The second type of measure can be exemplified by cultivation of catch crop or other land use measures such as increased area of grassland. The third type of measure consists of wetlands near the Baltic coast. For a detailed description of method and abatement measures in the static model, we refer to [26]. Based on data obtained from [26] ordinary least square estimator is applied for the estimation of coefficients in a quadratic cost function for nitrogen and phosphorus for each drainage basin, see Table B1 in Appendix B. This approach for deriving cost functions in each time period assumes that cost effective reductions of nitrogen and phosphorus are implemented in each drainage basin.

Finally there is a need for defining the time period when the targets in Table 1 are to be achieved, and to choose the level of the discount rate. Helcom BSAP suggests 2021 to be the deadline for implementation of nutrient load reductions. As was estimated from the "flushing out" scenario in [27], nutrient stocks in the entire sea have a response time scale of about 60-70 years. However, running the "flushing out scenario" in [27] indicate that even after over 130 years the sea did not come to a new nutrient balance with the nutrient loads reduced to "pre-industrial" levels [28]. Therefore, we assume that the nutrient concentration targets are to be achieved at the latest in year 2100 and then sustained for additional 70 years. With respect to the choice of discount rate, it can be noticed that there is no consensus in the large literature on the appropriate level of social discount rate. It is agreed that it is determined by a number of factors such as people's general time preferences, economic growth and utility from consumption. In practice, the long run economic growth rate is usually applied. This differs among the riparian countries, which would suggest different discount rates for the countries. However, this would create arbitrage possibilities of abatement between countries which is not consistent with a cost effective solution. We therefore apply a common discount rate of 0.03 which is in line with long run economic growth in several riparian countries.

3. Description of different climate change scenarios

We focus on climate change impacts and investigate their effects on cost effective solutions in isolation and in combination with future changes in nutrient loads due to development in demography and the agricultural sector. In the following, the derivations of impacts on nutrient loads from different climate change scenarios and changes in population and agriculture are presented.

3.1. Climate change and nutrient outflow

Climate change is expected to impact the hydrological water balance in the Baltic Sea region, leading to changes in river discharge to the sea. The general trend predicts an increase in precipitation and river outflow in the northern part of the drainage basin and a decrease in precipitation and river outflow in the southeast parts of the drainage basin [7].

To the best of our knowledge, there is no study analyzing the impact of climate change on nutrient outflow to all basins of the Baltic Sea, which is needed in this study. Furthermore, most studies calculate the impact of climate change only on nitrogen load to the Baltic Sea. Data from [5, 29] is therefore used to simulate the impact of climate change on nutrient outflow in the dynamic nutrient abatement model. Both [5] and [29] use the same four climate change scenarios, which are described in [7]. All of the four climate change scenarios are produced from a coupled regional atmosphere – Baltic Sea climate model, the so-called Rosby Centre Atmosphere Ocean Model (RCAO). Data from two different global general circulation models, from Hadley Centre, United Kingdom (HadAM3H) and Max Planck Institute for Metrology in Germany (ECHAM4/OPYC3), are used for setting the boundary conditions which drive the regional RCAO-model. Each of these model combinations applies two different CO_2 emission scenarios, high and low emissions, obtained from the Intergovernmental Panel on Climate Change (IPCC). The high emission scenario corresponds to a change in CO_2 equivalent content from the 1990 level of 353 ppm to 1143 ppm in the future. Correspondingly the low emission scenario implies an increase to 822 ppm [30]. This results in four different climate change scenarios with a high or a low future CO_2 level and with boundary conditions from one of two different global general circulation models. The time period for these scenarios stretches over a 30-year period 2071-2100 and is compared to a reference period of 1961-1990. These four climate change scenarios are labeled in the following way: "Climate change scenario 1"=RCAO-H/A2, "Climate change scenario 2"=RCAO-H/B2, "Climate change scenario 3"=RCAO-E/A2, "Climate change scenario 4"=RCAO-E/B2. Where RCAO=Rosby Centre, H=Hadley Centre, E=Max Planck Institute for metrology, A2=high emission scenario, B2=low emission scenario.

In [5] the predicted change in water discharges from [7] is used to model impacts of climate change on nitrogen outflow to five of the Baltic Sea marine basins; Bothnian Sea, Bothnian Bay, Baltic Proper, Gulf of Finland and Gulf of Riga. Corresponding nitrogen loads to the Danish strait and the Kattegat marine basins are obtained from [29]. However, neither [5] nor [29] model the impact of climate change on phosphorous loads to the Baltic Sea. This is made by [31] who shows that for the Finnish catchment Kokemäenjoki, climate change has an equally large impact on both nitrogen and phosphorous loads to the sea. Kokemäenjoki can be considered a representative catchment for the Bothnian Bay and Bothnian Sea basins with regard to climate and other characteristics [32, 7]. It is therefore assumed that the relationship between nitrogen and phosphorous loads due to climate change in Bothnian Bay and Bothnian Sea follow the same pattern as in Kokemäenjoki. In [33] the impacts of climate change on hydrology and nutrients in a Danish lowland river basin is analysed. Their results indicate that climate change reduces phosphorous loads to about 80 percent of the total change in nitrogen loads. It is assumed that this relationship applies also to the Baltic Proper, the Gulf of Finland, the Gulf of Riga, the Danish strait and the Kattegat.

Given these assumptions and the relationship between climate change and nutrient outflow as described in [5] and [29] the changes in nutrient loads under the four different scenarios are as presented in Table 2.

	Climate change-scenario1		Climate change-scenario 2		Climate change-scenario 3		Climate change-scenario 4	
	N	P	N	P	N	P	N	P
Bothnian Bay, Bothnian Sea	8	8	9	9	28	28	22	22
Baltic Proper	-32	-26	-17	-14	-61	-49	-19	-15
Gulf of Finland, Gulf of Riga	21	17	26	21	30	24	38	30
Danish Straits, Kattegat	11	9	15	12	33	26	31	25

Table 2. Changes in nitrogen, N, and phosphorus, P, loads to different marine basins from different climate models and assumed carbon dioxide emissions, in % from the reference case.

An interesting result in Table 2 is that climate change leads to calculated increases in nutrient outflows to all marine basins but the Baltic Proper. This is noteworthy since the size of the Baltic Proper basin and the stringency of the abatement goals for this basin (see Table 1) makes it important in any cost effective abatement scheme. As will be shown in Section 4 this turns out to have a major influence of climate change effects on cost efficient abatement solution.

3.3. Demographic change

Population growth and shifts towards coastal zones in the Baltic Sea catchment add to the impact from other drivers e.g. climate change [34]. In this respect changes in the costal-zone population have larger impacts on eutrophication since nutrient emission sources located further inland are affected by retention through plant assimilation, sedimentation and in the case of nitrogen denitrification. Demographic change scenarios would therefore ideally be based on demographic projections that take the distance to the sea into consideration. However to the best of our knowledge, projections of population change that make a distinction between the costal-zone and inland areas do not exist for the entire Baltic Sea drainage basin. Data are, however, available that allow for a distinction between rural and urban areas [35]. It is also possible to relate the population density for rural and urban areas respectively as functions of the distance to the Baltic Sea [36]. Using this functional relation the different impacts from demographic movements to the costal-zone and inland areas respectively on nutrient loads can be taken into consideration.

The projected impact on future nutrient loads from demographic change in the riparian countries is based on estimates of the population in 2008 and projections for 2050 for urban and rural areas [35]. Data on population density in rural and urban areas and their distance to the Baltic Sea are obtained from [36]. Under the assumption that the population density distribution as a function of distance to the Baltic Sea stays intact in the demographic projections, we can construct a demographic scenario where the different impact from demographic shifts to the costal-zone and inland areas respectively on nutrient loads to the Baltic Sea can be taken into consideration. The large coastal population of the Baltic Sea drainage basin [37] can thereby be factored into the analysis. In order to translate the demographic projections into changes in nutrient outflow we assume an annual production of 4,38 kg N/PE and 1,095 P/PE and unchanged shares of the population connected to sewage treatment [38]. It is further assumed that nutrient emissions from people living 1-10 km from the coast are not affected by nutrient retention. Emissions further inland is affected by retention according to ([26], Table A1). Table 3 presents the percentage increase/decrease in nitrogen and phosphorous load compared to the business as usual load presented in Table 1.

	Nitrogen	Phosphorous
Bothnian Bay	0.7	6.7
Bothnian Sea	0.9	7.7
Baltic Proper	-1.6	-10
Gulf of Finland	-1.9	-13.6
Gulf of Riga	-1.2	-1.2
Danish Straits	-2.1	0.2
Kattegat	0.7	7.2

Table 3. Changes in nitrogen and phosphorus loads to different marine basins from demographic change, in % from the nutrient loads in Table 1.

For most countries demographic change makes a larger impact on phosphorous loads to the Baltic Sea than on nitrogen loads. This is because sewage discharges, which depend on population size, contribute to approximately 50 percent of the total phosphorus load to the Baltic Sea compared to approximately 12 percent of the total nitrogen load [26]. The largest part of the population increase will take place in urban areas, close to the shore and is therefore not affected by retention. The largest decrease in population on the other hand takes place in rural areas with a larger part of the population living further from the coast and thus affected by retention. This will enhance the effect of demographic increases/decreases on nutrient outflow.

3.4. Future nutrient loads from agriculture

The future nutrient loads from agriculture are projected in [5] and based on assumed increase in consumption of animal protein for the year 2070, which is assumed to increase substantially in [5]. If this increased protein demand is met by domestic increase in animal production it would result in large increases in nutrient outflow to the Baltic Sea [39, 5]. The future increase in protein consumption is estimated in [5] based on the assumption that all countries in the Baltic Sea drainage basin will have protein consumption in 2070 equal to the mean of the EU-15 countries. Under this assumption time series for 1970-2003 is used to estimate protein consumption for the EU-15 countries and this relationship is then extended until 2070 (see [5] for details). Using the estimated increase in protein demand as a proxy for increased animal stock size they estimate consequential increase in nitrogen loads [5]. The impact on phosphorus load to the Baltic Sea due to structural change in the agricultural sector is however not included in [5]. Changes in phosphorous load needed to achieve the increase in total nitrogen load described in [5], have therefore been calculated based on constant proportions of nitrogen and phosphorus in livestock manure reported in ([26], Table B1, B2). This rough estimate of the phosphorous load together with the nitrogen estimations from [5] generates the increases in nitrogen and phosphorous due to increases in the animal production presented in Table 4.

	Nitrogen	Phosphorus
Bothnian Bay, Bothnian Sea	21	22
Baltic Proper	35	28
Gulf of Finland, Gulf of Riga	24	21
Danish Straits, Kattegat	51	20

Table 4. Changes in nitrogen and phosphorus loads to different marine basins from increased demand for animal protein in % from the nutrient loads in Table 1.

4. Cost effective achievement of the BSAP under different scenarios

Minimum costs are calculated for the impacts of the four different climate change scenarios presented in Section 3. For each climate change scenario the impact of demographic change and agricultural change occurring at the same time as climate change is also investigated. In Section 4.1 the impact of the four climate change scenarios on the cost effective implementation of the BSAP is presented in isolation. In Section 4.2 other future drivers of eutrophication are included in the form of changes in the demographic structure and structural changes in the agricultural sector in addition to climate change. These scenarios are then compared to the cost effective solution in the reference case (Section 2, Table 1). The GAMS Conopt2 solver is used for solving the problem [40]. In order to obtain tractable solutions, the entire period is divided into 30 periods where each period corresponds to 5 years. For all scenarios it is assumed that the full effect of the impact on future nutrient loads occurs from period one.

4.1. Climate change scenarios

As described in Section 3, climate change leads to increased nutrient outflow for all scenarios and all basins but the Baltic Proper. It might therefore be expected that climate change should lead to increased abatement costs. Inspection of Table 1 shows that the abatement targets are very stringent for phosphorus reductions to the Baltic Proper. This is the reason why total abatement costs decrease for all the climate change scenarios compared to the reference case except for scenario 4, see Figure 1.

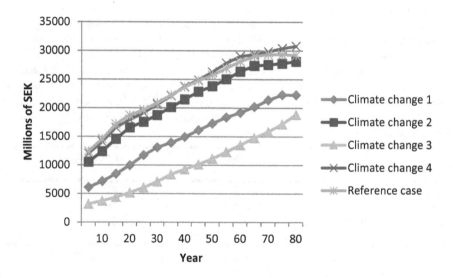

(SEK 1=€ 0,11; 2012-03-07)

Figure 1. Optimal paths of discounted abatement costs under different scenarios, Mill SEK/year.

The highest costs emerge under the fourth scenario with the lowest reductions in phosphorus loads to Baltic Proper and the largest increases in nutrients to the other basins. As expected, abatement is delayed as long as possible due to the discounting of future costs.

The abatement costs are also largest under Scenario 4 for most of the countries, see Figure 2. Climate change scenario 3, which results in the largest decrease in abatement costs, represents high future CO_2 emission scenario, as was shown in chapter three, and climate change scenario 4 represent a low future CO_2 emission scenario. This trend that abatement costs decrease with the severity of climate change is observed for all climate change models used in this paper. The reason is, as discussed in Section 2, that the cost reducing impact obtained by higher impact of abatement exceeds the cost increasing effect due to the need for large nutrient loads.

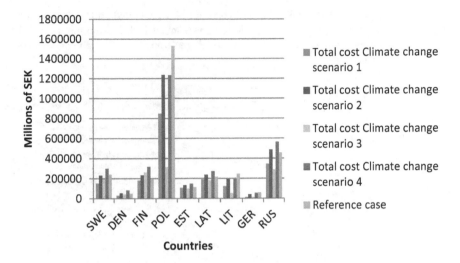

(SWE Sweden; DEN Denmark; FIN Finland; POL Poland; EST Estonia; LAT Latvia; LIT Lithuania; GER Germany; RUS Russia)

Figure 2. Total cost of abatement per country, under different climate change scenarios. (SEK 1=€ 0,11; 2012-03-07)

Common to all scenarios is the relative large abatement costs for Poland. This is because Poland is the largest polluter of nitrogen and phosphorous loads into the sea basins with the highest stringency in nutrient targets, the Baltic Proper. It is noteworthy that abatement cost for Poland decreases considerably under all climate change scenarios, in particular under climate change scenario 3, where the total abatement costs decrease by approximately 80 percent. Climate change also generates lower abatement costs for Germany and Lithuania but not as much as for Poland.

The impact of climate change on abatement costs creates a more diversified picture for the other riparian countries. For Russia, Latvia, Estonia, Denmark climate change scenario 2 and 4 lead to increases in abatement costs. For Sweden it is only climate change scenario 4 that implies larger abatement costs under a cost efficient implementation of the BSAP. For Finland all climate change scenarios except climate change scenario 1 leads to increases in abatement costs, this follows from the fact that Finland emits into the Bothnian Sea, Bothnian Bay and the Gulf of Finland where the largest increases in nutrient outflow due to climate change occur.

4.2. Combined scenarios

Climate change scenarios are combined with projections of development in the agricultural sector and demographic structure. Since the number of possible combinations is quite large, we focus on the climate change scenarios 3 and 4 which generated the lowest and highest abatement costs. The highest total abatement costs are now generated under climate change scenario 4 in combination with that on nutrient loads from agriculture, and the lowest costs are obtained under climate change scenario 3 in combination with demographic development, see Figure 3.

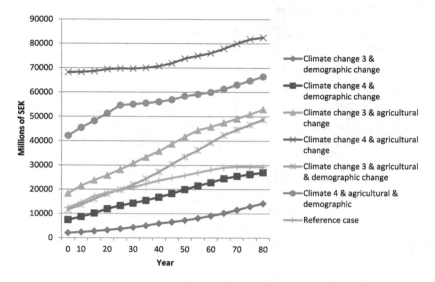

(SEK 1=€ 0,11; 2012-03-07)

Figure 3. Optimal paths of discounted abatement costs under different scenarios, Mill SEK/year

The inclusion of demographic change in the climate change scenarios results in a further decrease in abatement costs under all scenarios. This follows from the calculated decrease in population of approximately 15 percent for the entire Baltic Sea drainage basin (see Table B2 in the Appendix B). The decrease in total discounted abatement costs compared with the reference case then varies between 18 and 64 percent. The agricultural change scenario implies large increases in both nitrogen and phosphorous fluxes to the Baltic Sea. When agricultural change is simulated at the same time as climate change it results in an increase in total discounted abatement costs ranging between 65 and 200 percent depending on scenario.

Climate change and agricultural change are considered to be the major drivers of the future eutrophication of the Baltic Sea and demographic change will have an impact on future nutrient loads. It can therefore be argued that the most interesting scenario to consider is when all these drivers occur at the same time. When we include both agricultural change and demographic change in the climate change scenarios we observe an increase in abatement costs of 40-150 percent compared to the reference case. The agricultural change simulated here results in a very large increase in nutrient outflow to the Baltic Sea, and it might be argued that an increase of this magnitude is unlikely. It is however interesting that the major drivers work in different directions and that climate change and demographic change could counteract some of the increased abatement costs due to agricultural change.

Inspection of Table B3 in appendix B show that although total abatement cost increases under all climate change scenarios when both demographic change and agricultural change are included in the climate change scenarios, abatement cost decreases for Poland by 30 percent compared to the reference scenario under climate change scenario 3. It is thus possible that climate change could ease the cost burden for Poland, which carries the largest cost burden in all cost effective nutrient abatement schemes. Only Poland and Lithuania experience a decrease in abatement costs under climate change scenario 3, when agricultural change and demographic change are included in the climate change scenario analysis. For all other scenarios and countries abatement costs are larger than the reference scenario when agricultural change and demographic change are simulated at the same time as climate change.

5. Conclusions

The purpose of this paper has been to estimate impacts on costs for achieving the HELCOM targets for the Baltic Sea of different climate change scenarios in isolation and together with nutrient loads caused by future changes in agriculture and demography. Four different climate change scenarios, which are classified with respect to climate change model and projections of future carbon dioxide emissions, are investigated. The results indicate that impacts of climate change may facilitate the implementation of BSAP because of lower abatement costs. This occurs in spite of projected increases in nutrient outflow to the Baltic Sea for all marine basins but the Baltic Proper. The reason for this is the size of the Baltic Proper and the stringency in the abatement goals for this basin, in particular for phosphorus. An interesting feature of the scenario analysis is that abatement costs decrease with the severity of climate change, regardless of which climate change model is being used. These results are in line with the results found by [21] for the Baltic Sea sub-drainage basin Mälaren.

When we include both agricultural and demographic changes in the climate change scenarios we observe an increase in abatement costs corresponding to 40-150 percent compared to the reference case. In this scenario the major drivers of future nutrient loads to the Baltic Sea work in different directions. Climate and demographic changes both lead to lower total abatement costs while agricultural change leads to an increase in abatement costs. The increase in nutrient outflow from agricultural change is thus very large and the underlying assumptions that increased protein demand is met entirely by an increase in domestic animal production should be kept in mind. The magnitude of the increase in nutrient outflow caused by an increase in protein demand will be affected by this assumption and an increase in protein demand will to some extent be met by imported meat. The calculate cost increase from the agricultural change scenario can therefore be somewhat upward biased.

One should note that there are several limitations to the study; consideration has not been taken to the fact that the abatement targets as such might be altered by climate change. Another limitation is that uncertainty is not included in the climate change analysis, despite the fact that climate change most probably will lead to a change in the variability of nutrient loads to the Baltic Sea. An important factor in any future nutrient abatement scheme is development in abatement technology and, over a long time horizon, changes in preferences could also occur. Another limitation is the exclusion of response in the sea to climate change where e g. changes in water temperature and salinity level can affect the ecosystem in a manner that influence eutrophication and/or the environmental goal of clear water. These factors have not been considered in this study due to lack of data, but are important future developments of the analysis when data are available. Finally it is important to keep in mind that the results presented are scenarios and not predictions and should not be treated as such.

Appendix

A: Numerical discrete dynamic model and climate change scenarios

The numerical dynamic model is obtained from [20], with the inclusion of climate change parameters. In the following, we give a brief presentation of the model, and use it for analytical derivation of climate change impacts on optimal nutrient abatement.

Symbol	Explanation
$s, s=1,...,v$	drainage basin
$g, g=1,..,n$	country
$i, i=1,...k$	marine basin
$E, E=N,P$	nutrient loads, nitrogen (N) and phosphorus (P)
$t, t=0,..,T$	time period
I_t^{Eisg}	business as usual (BAU) nutrient load
M_t^{Eisg}	nutrient load
A_t^{Eisg}	nutrient abatement
$\phi^{HEsg}, 0 \leq \phi^{HEsg}$	proportional impact on BAU load under scenario H
L_t^{HEi}	nutrient load to a marine basin
$a^{Eji} = \dfrac{L^{Eji}}{L^{Ej}}$	transport coefficient in nutrient load from marine basin j to basin i
S_t^{HEi}	nutrient stock in a marine basin
$\alpha^{iE} \in (0,1]$	share of self cleaning of nutrient stock per period
W^{iE}	nutrient atom weight
K_T^{Ei}	nutrient concentration target in period T
$C^{sg}(A_t^{sg})$	abatement cost functions
$\rho_t = \dfrac{1}{(1+r)^t}$	discount factor with the discount rate r

Table A1. Definitions and explanation of symbols

Discharges from a specific sub-catchment into a marine basin in each time period is written as BAU loads minus abatement according to

$$M_t^{Eisg} = I_t^{Eisg} - f^{Eisg}(A_t^{sig})$$ (A1)

Our climate change quantification is assumed to have a multiplicative impact, on the reference loads, M_t^{Eisg}, so that nutrient loads in the scenario H is written as

$$M_t^{HEisg} = \varphi^{HEsg} M_t^{Eisg}$$ (A2)

The nutrient load to a marine basin is the sum of loads from its catchments and transports from other marine basins

$$L_t^{HEi} = \sum_s \sum_g M_t^{HEisg} + \sum_{j \neq i} \alpha^{Eji} L_t^{HEj}$$ (A3)

Stock dynamics of nutrient in a marine basin is

$$S_{t+1}^{HEi} = (1 - \alpha^{iE}) S_t^{HEi} + L_t^{HEi}$$ (A4)

$$S_0^{Ei} = S^{Ei}$$

The ecological targets are expressed in terms of nutrient concentrations as these are indicators of different types of ecological conditions e.g. [28]. The marine basin targets to be achieved in period T are then expressed as

$$((1 - \alpha^{iE}) S_Y^{HEi} + L_t^{HEi}) W^{Ei} \leq K_T^{Ei} \qquad \text{for} \quad i = 1, .., k \qquad E = N, P$$ (A5)

The decision problem is now specified as choosing the allocation of abatement among countries and time periods that minimises total control cost for achieving the targets defined by Eqs. (1)-(5), which is written as

$$\begin{array}{c} Min \\ A_t^{ig} \end{array} \qquad\qquad \sum_t \sum_s \sum_g \sum_E C^{ig}(A_t^{ig}) \rho_t$$ (A6)

s.t. (A1)-(A5)

The first-order conditions are obtained by formulating the Lagrangian which deliver

$$\rho_t \frac{\partial C^{ig}}{\partial A_t^{ig}} = \sum_j \sum_E \lambda_T^{jE} W^{jE} \sum_\tau (1 - \alpha^{jE})^{T-t+1} a^{ijE} \phi^{HEsg} \frac{\partial f^{Eigs}}{\partial A_t^{ig}}$$ (A7)

where λ_T^{iE} are the $k \times 2$ maximum number of Lagrange multipliers for the restrictions in k different marine basins with respect to two nutrient concentrations From (A1) to (A7) two counteracting impacts of climate change scenarios, i.e. ϕ^{HEsg}, can be identified; the effect on nutrient loads and associated impact on target achievement, and the effect of abatement measures on nutrient loads. The first effect can be seen from (A3) and (A5) where a higher proportional impact of the scenario on the reference nutrient load implies a larger nutrient load and accumulations. The second effect counteracts this cost increasing impact and is obtained from higher marginal effect of abatement on nutrient loads (see eq. A7). A larger impact from given marginal costs of abatement implies lower costs of achieving the targets. Another cost reducing effect is the delay of abatement which is increased since the impacts of later abatement is increased and can replace earlier abatement.

B: Tables and figures

Region	Nitrogen[1]:		Phosphorus[2]:		Coefficients in quadratic cost functions[3]	
	Kton	% of total	Kton	% of total	N	P
Denmark:		10.0		5		
Kattegat	36		0.8		14.15	4971
The Sound	30		0.9		4.71	2766
Finland:		6.8		9.5		
Bothnian Bay	16		1.5		8.79	4347
Bothnian Sea	18		1.2		8.21	2290
Gulf of Finland	11		0.5		7.78	2993
Germany:		10.7		1.5		
The Sound	23		0.3		8	61982
Baltic Proper	47		0.2		8.04	65525
Poland:		30.3		38.4		
Vistula	118		7.26		0.54	255
Oder	65		4.45		0.99	420
Polish coast	16		1.28		4.75	1483
Sweden:		14.2		11.0		
Bothnian Bay	9		0.95		64.93	10426
Bothnian Sea	18		1.14		24.99	2468
Baltic Proper	26		0.81		6.49	3230
The Sound	6		0.1		6.38	13118
Kattegat	34		0.72		2.95	6712
Estonia:		3.7		3.6		
Baltic Proper	1		0.02		18.77	20227
Gulf of Riga	10		0.25		10.03	9432
Gulf of Finland	13		0.93		1.33	2160
Latvia:		9.0		6.2		
Baltic Proper	8		0.25		22.27	5522
Gulf of Riga	51		1.84		4.93	1635
Lithuania	42	6.4	2.35	7.0	39.55	1268
Russia:		9.0		18.0		
Baltic Proper	10		1.19		43.62	5846
Gulf of Finland	49		4.90		4.68	734
	657	100	33.8	100		

1.Tables B1 and B3 in [20]; 2. Table B2 in [20]; 3 $TC=a(N^{Bau}-N)^2+ b(P^{Bau}-P)^2$ where TC is total cost, N^{Bau} and P^{Bau} in the reference case, and N and P are the optimal loads for achieving nutrient concentration targets [26].

Table B1. BAU nitrogen, N, and phosphorus, P, loads from different drainage basins of the Baltic Sea, kton and in % of total loads in the reference case, estimated coefficients in nutrient abatement cost functions

	Total population 2008, thousand	Total population 2050, thousand	Demographic change, thousand	%
Estonia	1341	1233	-108	-8
Finland	5304	5445	141	2,7
Latvia	2259	1854	-405	-18
Lithuania	3321	2579	-742	-22
Poland	38104	32013	-6091	-16
Russian federation	141394	116097	-25297	-18
Sweden	9205	10571	1366	15
Germany	82264	70504	-11760	-14
Denmark	5458	5551	93	2
Total	287309	244614	-42695	-15

Table B2. Demographic change in countries of the Baltic Sea drainage basin.

	Climate change scenario 3, & agricultural & demographic change	Climatechange scenario 4, & agricultural & demographic change	Reference case
Sweden	773325	1093329	241085
Denmark	369721	467326	48896
Finland	671922	901168	194423
Poland	1058937	2257641	1529647
Estland	183267	243281	114808
Latvia	353262	644229	216339
Lithuania	142966	439524	245249
Germany	144004	471943	59343
Russia	626599	1460392	456101
Total abatement cost	4324003	7978833	3105891

Table B3. Total abatement cost per country for climate change scenario 3 and 4, when demographic change and agricultural change is simulated at the same time. Millions of SEK.

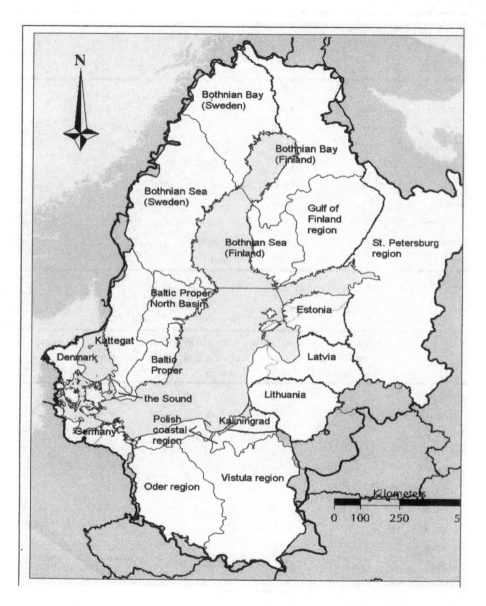

Figure B1. Drainage basins of the Baltic Sea (originally from [41]). (Drainage basins in Denmark (2). Germany (2). Latvia (2). and Estonia (3) are not provided with names. but are delineated only by fine lines)

Author details

Martin Lindkvist, Ing-Marie Gren and Katarina Elofsson
Department of Economics,
Swedish University of Agricultural Sciences,
Uppsala,
Sweden

6. References

[1] Gilbert P M. Eutrophication and Harmful Algal Blooms: A Complex Global Issue, Examples from the Arabian Seas and including Kuwait Bay and an introduction to the Global Ecology and Oceanography of Harmful Algal Blooms (GEOHAB) Programme. International Journal of Oceans and Oceanography 2007; 2 (1) 157-169.

[2] Heisler J, Glibert P M, Burkholder J M,. Anderson D M, Cochlan W, Dennison W C, Dortch Q, Goble C J,. Heil C A, Humphries E, Lewitus A, Magnien R, Marshall H G, Sellner K, Stockwell D A, Stoecker D K, Suddleson M, Eutrophication and harmful algal blooms: A scientific consensus. Harmful Algae 2008; 8 3–13

[3] Elmgren R., Larsson U., Eutrophication in the Baltic Sea area. Integrated coastal management issues. In: B. von Bodungen and R.K. Turner (Eds.). Science and integrated coastal management. Dahlem University Press, Berlin; 2001: p15-35

[4] Conley D J, Björck S, Bonsdorff E, Carstensen J, Destouni G, Gustafsson B, Hietanen S. Kortekaas M, Kuosa H, Meier M, Müller-Karulis B, Nordberg K, Norkko A, Nürnberg G, Pitkänen H, Rabalais N, Rosenberg R, Savchuk O, Slomp C, Voss M. Wulff F, Zillén L. Hypoxia-related processes in the Baltic Sea, Environmental Science and Technology, 2009; 43 3412-.3420

[5] Eriksson-Hägg H, Humborg C, Mörth C-M, Medina M R, Wulff F. Scenario analysis on protein consumption and climate change effects on riverine N export to the Baltic Sea. Environmental Science and Technology 2010; 44(7) 2379-2385

[6] HELCOM, Climate Change in the Baltic Sea Area – HELCOM Thematic Assessment in 2007 Baltic Sea Environ. Proc. No. 111; 2007

[7] Graham L. Climate Change Effects in River Flow to the Baltic Sea. Ambio 2007; 33 235-241.

[8] Gren, I-M, Elofsson K, Jannke P. Cost effective nutrient reductions to the Baltic Sea. Environmental and Resource Economics 1997; 10(4) 341-362.

[9] Elofsson K. Cost effective reductions in the agricultural load of nitrogen to the Baltic Sea. In Bohman. M. Brännlund. R. Kriström. B. (eds.). Topics in environmental Economics. Amsterdam, the Netherlands: Kluwer Academic Publisher; 1999

[10] Elofsson K. Cost effective control of interdependent water pollutants. Environmental Management 2006; 37(1) 54-68.

[11] Elofsson K. Cost uncertainty and unilateral abatement. Environmental and Resources Economics 2007; 36 (2) 143-162.

[12] Gren I-M. International versus national actions against pollution of the Baltic Sea. Environmental and Resource Economics 2001; 20(1) 41-59.

[13] Gren I-M. Mitigation and adaptation policies for stochastic water pollution: An application to the Baltic Sea. Ecological Economics 2008; 66(2-3) 337-347.

[14] Ollikainen M, Honkatukla J. Towards efficient pollution control in the Baltic Sea: An anatomy of current failure with suggestions for change. Ambio 2001; 4779-93.

[15] Hart R, Brady M. Nitrogen in the Baltic Sea – Policy implication of stock effects. Journal of Environmental Management 2002; 66(1) 91-103

[16] Hart R. Dynamic pollution control. Ecological Economics 2003; 47 79-93.

[17] Gren I-M, Wulff F. Cost effective management of polluted coupled heterogeneous marine basins in the Baltic Sea. Regional Environmental Change 2004 ; 4 1-24.

[18] Laukkanen M, Huhtala A. Optimal management of a eutrophied coastal ecosystem: balancing agricultural and municipal abatement measures. Environmental and Resource Economics 2008; 39 139-159.

[19] Laukkanen M, Ekholm P, Huhtala A, Pitkänen H, Kiirikki M, Rantanen P, Inkata A . Integrating ecological and economic modelling of eutrophication: Toward optimal solutions for a coastal area suffering from sediment release of phosphorus. Ambio 2009; 38(4) 225-235.

[20] Gren I-M, Savchuck O, Jansson T. Dynamic cost effective mitigation of eutrophication in the Baltic Sea with coupled hetereogeneous marine basins. Marine Resource Economics; in revision 2012

[21] Gren I-M. Climate change and the Water Framework Directive: Cost effectiveness and policy design for water management in the Swedish Mälar region. Climatic Change 2010; 100(3) 463-484.

[22] Helcom HELCOM Baltic Sea Action Plan. Helsinki Commission, Helsinki, Finland. http://www.helcom.fi/BSAP (accessed 18.04.10)

[23] Backer H, Leppänen J-M, Brusendorff AC, Forsius K, Stankiewicz M, Mehtonen J, Pyhälä M, Laamanen M, Paulomäki H, Vlasov N, Haaranen T. HELCOM Baltic Sea Action Plan – a regional programme of measures for the marine environment based on the Ecosystem Approach. Marine Pollution Bulletin 2010; 60(5) 642-649

[24] Savchuk O. Resolving the Baltic Sea into seven subbasins: N and P budgets for 1991-1999. Journal of Marine Systems 2005; 56 1-15.

[25] Griffin J H. Joint production technology: the case of petrochemicals. Econometrica 1978; 46(2) 379 – 396.

[26] Gren I-M, Lindqvist M, Jonzon Y. Calculation of costs for nutrient reductions to the Baltic Sea – technical report. Working paper no. 2008-1. Department of Economics, SLU, Uppsala, 2008.

[27] Savchuk O P, Wulff F. Modeling the Baltic Sea eutrophication in a decision support system. Ambio 2007; 36(2-3) 141 – 148.

[28] Savchuk OP, Wulff F. Long-term modelling of large-scale nutrient cycles in the entire Baltic Sea. Hydrobiologia 2009; 629 209-224.

[29] Arheimer B, Andréasson J, Fogelberg S, Johansson H, Pers C, Persson K. Climate change impact on water quality: model results from southern Sweden. Ambio 2005; 34(7) 559-566.

[30] Nakic'enovic', N. et al. Special Report on Emissions Scenarios (SRES); Intergovernmental Panel on Climate Change (IPCC), 599 pp, 2000.

[31] Kaipainen H, Bilaletdin Ä, Frisk T, Paananen A. The impact of climate change on nutrient flows in the catchment of river Kokemäenjoki. BALWOIS, Ohrid, Republic of Macedonia – 27, 31 May, 2008

[32] Humborg C, Danielsson Å, Sjöberg B, Green M. Nutrient land–sea fluxes in oligothrophic and pristine estuaries of the Gulf of Bothnia, Baltic Sea. Estuarine, Coastal and Shelf Science 2003; 56 781–793

[33] Andersen H E, Kronvang B, Larsen S E, Hoffmann CC, Jensen T, Rasmussen S. Climate-change impacts on hydrology and nutrients in a Danish lowland river basin. Science of the Total Environment 2006; 365 223-237.

[34] Rabalais N N, Turner R E, Di'az R J, Justic' D. Global change and eutrophication of coastal waters. – ICES Journal of Marine Science 2009; 66 1528–1537

[35] FAO Faostat: Projected change in rural and urban population.
http://faostat.fao.org/site/550/DesktopDefault.aspx?PageID=550
(accessed 2010-03-25)

[36] Baltic GIS Statistics Land Cover & Population Statistics, Table 3.
http://www.grida.no/baltic/htmls/stat.htm
(accessed 2010-03-25).

[37] Eurostat- Unit E1, Farms, Agro-environment and rural Development. Nearly half of the population of EU countries with a sea border is located in coastal regions. Statistics in focus 2009; 47.

[38] Shou J S, Neye ST, Lundhede T, Martinsen L, Hasler B. Modelling cost-efficient reductions of nutrient loads to the Baltic Sea. NERI technical report No. 592. Ministry of the Environment, Copenhagen, Denmark; 2006

[39] Humborg C , Mörth C-M, Sundbom M, Wulff F. Riverine transport of biogenic elements to the Baltic Sea – past and possible future perspectives. Hydrology and Earth System Sciences 2007; 11 1593-1607

[40] Brooke A, Kendrick D, Meeraus A. Gams – a user's guide. San Francisco: The Scientific Press, USA; 1998.

Permissions

The contributors of this book come from diverse backgrounds, making this book a truly international effort. This book will bring forth new frontiers with its revolutionizing research information and detailed analysis of the nascent developments around the world.

We would like to thank Prof. (Dr.) Bharat Raj Singh, for lending his expertise to make the book truly unique. He has played a crucial role in the development of this book. Without his invaluable contribution this book wouldn't have been possible. He has made vital efforts to compile up to date information on the varied aspects of this subject to make this book a valuable addition to the collection of many professionals and students.

This book was conceptualized with the vision of imparting up-to-date information and advanced data in this field. To ensure the same, a matchless editorial board was set up. Every individual on the board went through rigorous rounds of assessment to prove their worth. After which they invested a large part of their time researching and compiling the most relevant data for our readers. Conferences and sessions were held from time to time between the editorial board and the contributing authors to present the data in the most comprehensible form. The editorial team has worked tirelessly to provide valuable and valid information to help people across the globe.

Every chapter published in this book has been scrutinized by our experts. Their significance has been extensively debated. The topics covered herein carry significant findings which will fuel the growth of the discipline. They may even be implemented as practical applications or may be referred to as a beginning point for another development. Chapters in this book were first published by InTech; hereby published with permission under the Creative Commons Attribution License or equivalent.

The editorial board has been involved in producing this book since its inception. They have spent rigorous hours researching and exploring the diverse topics which have resulted in the successful publishing of this book. They have passed on their knowledge of decades through this book. To expedite this challenging task, the publisher supported the team at every step. A small team of assistant editors was also appointed to further simplify the editing procedure and attain best results for the readers.

Our editorial team has been hand-picked from every corner of the world. Their multi-ethnicity adds dynamic inputs to the discussions which result in innovative outcomes. These outcomes are then further discussed with the researchers and contributors who give their valuable feedback and opinion regarding the same. The feedback is then collaborated with the researches and they are edited in a comprehensive manner to aid the understanding of the subject.

Apart from the editorial board, the designing team has also invested a significant amount of their time in understanding the subject and creating the most relevant covers. They scrutinized every image to scout for the most suitable representation of the subject and create an appropriate cover for the book.

The publishing team has been involved in this book since its early stages. They were actively engaged in every process, be it collecting the data, connecting with the contributors or procuring relevant information. The team has been an ardent support to the editorial, designing and production team. Their endless efforts to recruit the best for this project, has resulted in the accomplishment of this book. They are a veteran in the field of academics and their pool of knowledge is as vast as their experience in printing. Their expertise and guidance has proved useful at every step. Their uncompromising quality standards have made this book an exceptional effort. Their encouragement from time to time has been an inspiration for everyone.

The publisher and the editorial board hope that this book will prove to be a valuable piece of knowledge for researchers, students, practitioners and scholars across the globe.

List of Contributors

P. J. M. Cooper
School of Agriculture, Policy and Development, University of Reading, UK

R. D. Stern
Statistical Services Centre, University of Reading, UK

M. Noguer and J. M. Gathenya
Walker Institute for Climate System Research, University of Reading, UK

Nataliya Moskalenko
Earth Cryosphere Institute SB RAS, Russia

Hui Lu
Ministry of Education Key Laboratory for Earth System Modeling, and Center for Earth System Science, Tsinghua University, Beijing, China

Toshio Koike, Tetsu Ohta and Katsunori Tamagawa
The Department of Civil Engineering, The University of Tokyo, Tokyo, Japan

Hideyuki Fujii
Earth Observation Research Center, Japan Aerospace Exploration Agency, Ibaraki, Japan

David Kuria
Geomatic Engineering and Geospatial Information Science Department, Kimathi University College of Technology, Kenya

Tony Prato
University of Missouri, USA

Zeyuan Qiu
New Jersey Institute of Technology, USA

S.C. Nwanya
Department of Mechanical Engineering, University of Nigeria, Nsukka, Nigeria

Youmin Tang
Environmental Science and Engineering, University of Northern British, Columbia, Prince George, Canada
State Key Laboratory of Satellite Ocean Environment Dynamics, Second Institute of Oceanography, State Oceanic Administration, Hangzhou, P. R. China

Dake Chen
State Key Laboratory of Satellite Ocean Environment Dynamics, Second Institute of Oceanography, State Oceanic Administration, Hangzhou, P. R. China
Lamont-Doherty Earth Observatory, Columbia University, Palisades, NY, USA

Dejian Yang
School of Atmospheric Sciences, Nanjing University

Tao Lian
State Key Laboratory of Satellite Ocean Environment Dynamics, Second Institute of Oceanography, State Oceanic Administration, Hangzhou, P. R. China

Jaroslav Solár
Institute of High Mountain Biology, University of Zilina, Slovakia

Erik Vanem, Bent Natvig and Arne Bang Huseby
University of Oslo, Norway

Elzbieta M. Bitner-Gregersen
Det Norske Veritas, Norway

Martin Lindkvist, Ing-Marie Gren and Katarina Elofsson
Department of Economics, Swedish University of Agricultural Sciences, Uppsala, Sweden